W9-ADN-614

WITHDRAWN
L. R. COLLEGE LIBRARY

WILL
L.R. COM...

Exploring the Universe

Exploring the Universe

Essays on Science and Technology

Edited by
P. DAY

Oxford   New York   Tokyo
OXFORD UNIVERSITY PRESS
THE ROYAL INSTITUTION
1997

CARL A. RUDISILL LIBRARY
LENOIR-RHYNE COLLEGE

Q
158.5
.E 97
1997

7ab.1999

Oxford University Press, Great Clarendon Street, Oxford OX2 6DP

Oxford   New York
Athens   Auckland   Bangkok   Bogota   Bombay   Buenos Aires
Calcutta   Cape Town   Dar es Salaam   Delhi   Florence   Hong Kong
Istanbul   Karachi   Kuala Lumpur   Madras   Madrid   Melbourne
Mexico City   Nairobi   Paris   Singapore   Taipei   Tokyo   Toronto
and associated companies in
Berlin   Ibadan

Oxford is a trade mark of Oxford University Press

Published in the United States
by Oxford University Press Inc., New York

© The Royal Institution of Great Britain, 1997

*All rights reserved. No part of this publication may be
reproduced, stored in a retrieval system, or transmitted, in any
form or by any means, without the prior permission in writing of Oxford
University Press. Within the UK, exceptions are allowed in respect of any
fair dealing for the purpose of research or private study, or criticism or
review, as permitted under the Copyright, Designs and Patents Act, 1988, or
in the case of reprographic reproduction in accordance with the terms of
licences issued by the Copyright Licensing Agency. Enquiries concerning
reproduction outside those terms and in other countries should be sent to
the Rights Department, Oxford University Press, at the address above.*

*This book is sold subject to the condition that it shall not,
by way of trade or otherwise, be lent, re-sold, hired out, or otherwise
circulated without the publisher's prior consent in any form of binding
or cover other than that in which it is published and without a similar
condition including this condition being imposed
on the subsequent purchaser.*

A catalogue record for this book is available from the British Library

Library of Congress Cataloging in Publication Data
Data available

ISBN 0 19 850085 8

Typeset by Footnote Graphics, Warminster, Wiltshire
Printed in Great Britain by
Bookcraft Ltd., Midsomer Norton, Avon

PREFACE

Since 1826, when they were started by Michael Faraday, the Friday Evening Discourse Programme at the Royal Institution has been bringing contemporary science and technology to a general audience through the medium of illustrated lectures by noted authorities, among whom Faraday himself was one of the most gifted. The complete series of texts is made available to members of the Royal Institution as one of the privileges of membership, but in recent years a selection has been published separately by Oxford University Press for the public. It is pleasure to introduce the present selection, comprising entertaining and authoritative accounts of topics from the whole gamut of science, from visual perception through radioactivity to the constituents of interstellar space. I hope that readers who find the present volume interesting will be encouraged to become members of the Royal Institution themselves, and receive the full collection.

London   P. D.
March 1997

CONTENTS

PLATES

Plates 1–13 appear between pages 84 and 85.

1. The responses of a cell in area V4. This cell responds best to a blue square on a white backgrond.
2. *Red Square* by Kazemir Malevich.
3. PET scan for [^{11}C]diprenorphine, a ligand for the opiate system, showing uptake at specific receptor sites in the brain. (From a photograph kindly provided by the MRC Clinical Sciences Centre.)
4. Successive Hubble Space Telescope pictures of Jupiter showing the influence of winds on some Comet Shoemaker–Levy 9 impact sites over a time of 5 days. (Reproduced with permission of the Space Telescope Science Institute (STScI) and NASA.)
5. Hubble Space Telescope picture of the planetary nebula NGC 6543, nicknamed the Cat's Eye Nebula. (Reproduced with permission of STScI and NASA.)
6. Region of the supernova remnant known as the Cygnus Loop imaged with the Hubble Space Telescope. The stepped outline of this and other pictures in this article comes from the way the images are assembled. (Reproduced with permission of STScI and NASA.)
7. Hubble Space Telescope picture of the supermassive, eruptively unstable star Eta Carinae showing billowing lobes of gas and dust. (Reproduced with permission of STScI and NASA.)
8. Hubble Space Telescope picture of a region in the Orion Nebula. (Reproduced with permission of STScI and NASA.)
9. Enlarged view of a region in Plate 8 containing young stars with protoplanetary disks of gas and dust. (Reproduced with permission of STScI and NASA.)
10. Hubble Space Telescope picture of the face-on spiral galaxy M100 in the Virgo Cluster of galaxies. (Reproduced with permission of STScI and NASA.)
11. The deepest, most detailed optical view of the Universe yet obtained, showing galaxies in the early Universe. This Hubble Space Telescope picture covers a speck of sky only one-thirtieth the diameter of the full Moon. (Reproduced with permission of STScI and NASA.)
12. An early distillation apparatus.
13. The kudzu plant.

CONTRIBUTORS

Alec Boksenberg
Institute of Astronomy,
University of Cambridge,
Madingley Road,
Cambridge CB3 0HA

Peter Christmas
12 Bray Road,
Stoke D'Abernon,
Cobham,
Surrey KT11 3HZ

Chris Elliott
Magalee,
Moon Hall Road,
Ewhurst,
Surrey GU6 7NP

John Mann
Professor of Organic Chemistry,
The University of Reading,
Whiteknights,
Reading RG6 6AD

James McQuaid
Chief Scientist,
Health and Safety Executive,
Rose Court,
2 Southwark Bridge,
London SE1 9HS

Bert Vallee
Edgar M. Bronfman, Sr,
Distinguished Senior Professor of
Biochemistry,
Harvard Medical School,
Boston, MA 02115,
USA

Semir Zeki
Professor of Neurobiology,
Department of Anatomy,
University College London,
London WC1E 6BT

Aphrodisiacs, psychedelics, and the elusive magic bullet

JOHN MANN

Human cultures have always experimented with the plants and animals with which they coexisted: Poisonous extracts were often employed as agents of warfare or for hunting, execution, and suicide. Those products that were stimulants or produced visual (and other) illusions were used to alleviate fatigue or for magico-religious purposes. Finally, the extracts that possessed apparent medicinal properties were the most highly prized. The village shamans became adept at using these natural extracts for murder, magic, and medicine and it is the knowledge in this last area that has been so useful to the pharmaceutical industry in its quest for new drugs. In this account the two extremes of pharmacological activity will be explored through an examination of those extracts that (apparently) enhanced sexual potency or produced hallucinations, and those that provided the inspiration for studies aimed at the development of new drugs, especially those with specific modes of activity—the 'magic bullets'.

Aphrodisiacs and psychedelics

The male Indian rhinoceros is reputed to spend as long as one hour on the sex act and to ejaculate as many as ten times during this period. Not surprisingly, such libidinous activity excited the interest of human males and for centuries various parts of the rhinoceros, most notably the horn, have been prescribed as agents for enhancing sexual potency. A few grams of powdered horn costs about £20, but since the horn is made from the structural protein keratin, this costly commodity is likely to be as efficacious as human nail clippings, which are also made of keratin.

Several members of the insect family Meloides, the so-called blister beetles, have somewhat greater notoriety. They produce a chemical compound called cantharidine which has potent vesicant properties and, if taken internally, not only causes serious damage to the gastrointestinal tract but also irritates the urethra, thus producing prolonged excitation of erectile tissue. The resultant state of priapism may mimic the effects of extreme sexual excitement, but the toxic and irritant effects of this natural product far outweigh any apparent aphrodisiac effect.

A similar though more benign effect is produced by the alkaloid yohimbine from the African plant *Corynanthe yohimbe*. This compound increases blood flow to the sexual organs and lead to excitation of erectile tissue in both men and women. A number of synthetic drugs have a similar affect: Captagon is a stimulant much prized by Arabian men since it delays ejaculation; and gamma-hydroxybutyric acid appears to reduce inhibitions and induce sexual stimulation.

Fig. 1 *The Absinthe Drinker* by Edouard Manet (1859).

Pride of place must, however, be reserved for the nineteenth century liqueur absinthe, since more poets, writers, and artisans extolled the mind-expanding and aphrodisiac properties of this drink than any other preparation. The recipe for absinthe is said to have originated with a French doctor called Pierre Ordinaire in around 1792. Quantities of the plant *Artemisia absinthium* (wormwood), together with similar amounts of fennel and anise, were steeped in strong grape brandy for a period prior to distillation. The resultant bitter, green liqueur was usually diluted with water and sweetened with sugar to produce an opalescent brew of considerable potency. Its supposed aphrodisiac properties provided an early allure, but its popularity arose primarily from its use by French soldiers during their campaign in North Africa in the 1840s. It is likely that they valued its mild antiseptic and anthelmintic properties that lessened the risk of dysentry and infestation by intestinal worms. The period following the Franco–Prussian war (1870–71) was a time of peace and enlightenment and *l'heure verte* became firmly established especially amongst the artisans and literati. Many of the contemporary artists depicted absinthe including Manet, in *The Absinthe Drinker* (1859) (Fig.1) and Degas (1876) and Van Gogh (1887), both in works entitled *L'Absinthe*. Writers extolled the virtues of *la fée verte* (the 'green fairy'):

> *On boit sans perdre la raison*
> *Cette boisson si délectable,*
> *Elle est l'ami de Cupidon.*
> *Auprès du sexe on est aimable,*
> *On voit en rose l'avenir,*
> *Elle vous fait aimer la vie*
> *Et le plus mauvais souvenir*
> *Grace à ses charmes vite s'oublie!*
> Arthur Rimbaud

And also its darker side:

> *Je suis la Fée Verte,*
> *Je suis la ruine et la douleur,*
> *Je suis la honte,*
> *Je suis le deshonneur,*
> *Je suis la mort,*
> *Je suis l'Absinthe.*
> Arthur Rimbaud

Its excitation of the mind and its aphrodisiac properties were more significant than these problems, at least for the moderate drinker, and by 1913 the factories of Pernod Fils and other companies were producing more than 10 million gallons of absinthe per annum. For those who drank immoderately, the condition of absinthism was likely to develop. This was manifested initially by visual and auditory hallucinations, together with hyperexcitability; over a period of years the absinthists became pallid and enfeebled and usually exhibited symptoms of mental derangement. This toxicity was ascribed to the presence of the natural product thujone, which is the major constituent of *Artemisia absinthium*, and was probably present to the extent of around 4 mg per glass of absinthe, making it the major chemical species present other than ethanol. The presence of methanol and copper salts, which were often added to improve the green colour, were overlooked and, although efforts were made to remove the thujone, the liqueur was eventually banned in France in 1915. The original recipe of Pierre Ordinaire still survives in the form of the aperitifs Pernod and Ricard though these do not contain *Artemisia absinthium*.

As for the pharmacological action of absinthe, there were suggestions in the 1970s that thujone and tetrahydrocannabinol (THC)—the major psychoactive constituent of cannabis—had sufficient structural similarity to share a receptor site in the brain. The excitatory properties of both compounds could thus perhaps be explained. With the recent tentative identification of the cannabinoid receptor and of its natural substrate, christened anandamide, it should soon be possible to test this theory. Our computer modelling studies at Reading (see the structures shown in Fig. 2) have indeed demonstrated that there are structural similarities between THC and anandamide, though their similarity to thujone is more tenuous. We are presently preparing isotopically labelled samples of all three compounds to check whether they bind to the anandamide receptor, so perhaps we shall soon be able to throw more light upon the pharmacological basis of the alleged stimulant and aphrodisiac activity of absinthe.

There is much less uncertainty about the pharmacology of the various psychomimetic plant extracts. For example, there is no doubt that the hallucinogenic properties of the Aztec magical preparation *ololuiqui* (from the seeds of *Rivea corymbosa*) were due to the constituent ergot alkaloids that are closely related in structure to lysergic acid diethylamide (LSD). The Spanish adventurer and physician Francisco Hernandez said of *ololuiqui*: 'When the priests wanted to commune with their Gods ... they ate the seeds and a thousand visions and satanic hallucinations appeared to them'. In the late 1950s the ethnobotanist Richard Wasson sent some of these seeds to Albert Hofmann, discoverer of LSD, and encouraged him to prepare his own extract to compare its psychomimetic effects with those he had experienced with LSD. Indeed, he found the experience was similar to that elicited by LSD, though *ololuiqui* was much less potent. This

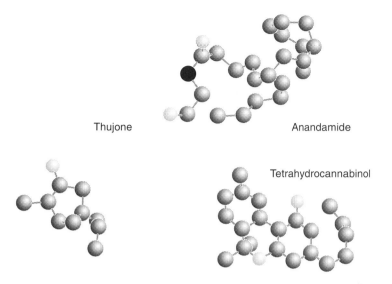

Thujone   Anandamide

Tetrahydrocannabinol

Fig. 2 Comparison of the structures of thujone, anandamide, and tetrahydrocannabinol. Here and in subsequent structures,

 represents an oxygen atom

represents a carbon atom

represents a nitrogen atom

fact is not surprising given the similarity in structure between LSD and the major ergot alkaloid from *R. corymbosa* (see Fig. 3), and both probably exert their effects through interaction with 5-hydroxytryptamine receptors in the brain. 5-Hydroxytryptamine (5-HT) is a major human neurotransmitter and is implicated in the control of sleep states, satiation, and mood; the 5-HT structure is clearly discernible within the larger structures of LSD and its South American counterpart. Aberrant excitation of 5-HT receptors is likely to provoke some kind of psychomimetic activity, and a whole range of other South American plants yield hallucinogenic snuffs and other preparation whose psychoactive constituents (for example dimethyltryptamine) resemble 5-HT even more closely (see Fig. 3).

However, one does not need to travel as far as South America to find plants and fungi that contain hallucinogens, since our own *Amanita*

lysergic acid diethylamide (LSD)

5-hydroxytryptamine (5-HT)

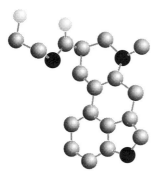

'ololuiqui'

dimethyltryptamine (DMT)

Fig. 3 The structures of (left) lysergic acid diethylamide (LSD) and the Aztec magical preparation *ololuiqui*, and (right) 5-hydroxytryptamine and dimethyltryptamine.

muscaria, or fly agaric, has a long association with folklore and magic. The descriptions given by the eighteenth century explorers Oliver Goldsmith and Georg Steller leave little doubt of the efficacy of the fungi:

> *Those who are rich among them, lay up large provisions of these mushrooms, for the winter. When they make a feast, they pour water upon some of the mushrooms, and boil them. They then drink the liquor, which intoxicates them. The poorer sort ... watch the opportunity of the guests coming down to make water; and then hold a wooden bowl to receive the urine, which they drink off greedily ... and by this way they also get drunk.*
> Georg Steller

> *The fly agarics are dried, then eaten in large pieces ... After about half an hour the person becomes completely intoxicated and experiences extraordinary visions. Those who cannot*

*afford the fairly high price [of the mushrooms] drink the urine
of those who have eaten it, whereupon they become as intoxi-
cated, if not more so. The urine seems to be more powerful than
the mushroom, and its effect may last through the fourth or
fifth man.*

Oliver Goldsmith

The urine drinking deserves some comment and can probably be
explained by consideration of the metabolism of the major constituents
of the fungi. These psychoactive constituents are ibotenic acid and the
more potent muscimol (see Fig. 4); they appear to exert their activity by
binding to the brain receptors for the natural neurotransmitter gamma-
aminobutyric acid (GABA). This is one of the most important inhibitory
neurotransmitters in the brain and the two hallucinogens act as GABA
agonists—that is they cause excitation at the receptors. It is apparent
from Fig. 4 that the three structures share a common backbone so competition for the receptor seems reasonable. After consumption, ibotenic
acid can be metabolized to the more potent muscimol through loss of
carbon dioxide (CO_2), so the urine of the consumer is likely to contain
rather more muscimol than in the original mushroom. The urine
drinkers thus had the better part of the deal!

While none of these hallucinogens has clinical utility, an understanding of their interactions with brain receptors should help to provide a
greater insight into the normal functioning of the brain, and perhaps suggest ways in which neurological disease can be treated. Indeed, the new
anti-migraine drug Sumatriptan and the anti-emetic agent Ondansetron
were both designed using knowledge about 5-HT receptors. The former
drug exerts its actions on 5-HT receptors in the brain and the latter acts
upon receptors found in the gut.

Finally, no account of mind-altering drugs would be complete without
some mention of the mandrake, *Mandragora officinarum*. This 'human-
shaped' root has been associated with murder, magic, and medicine
since biblical times. Greek physicians used sublethal quantities of root
extracts to produce a numbing effect in their surgical patients, while
mediaeval witches (Fig. 5) anointed themselves 'under the arms and in
other hairy places' (to ensure percutaneous uptake) and attained a state
of euphoria and disorientation—hence the flying sensation! The arch-
poisoners of ancient and more recent times used larger doses of the root
extract to despatch their victims.

The major active constituent, scopolamine or hyoscine, is an important
ingredient of the popular surgical premedication Omnopon/scopolamine
—a mixture of scopolamine and opium alkaloids. This is given to produce a state of relaxation in the patient but also serves to dry up secre-

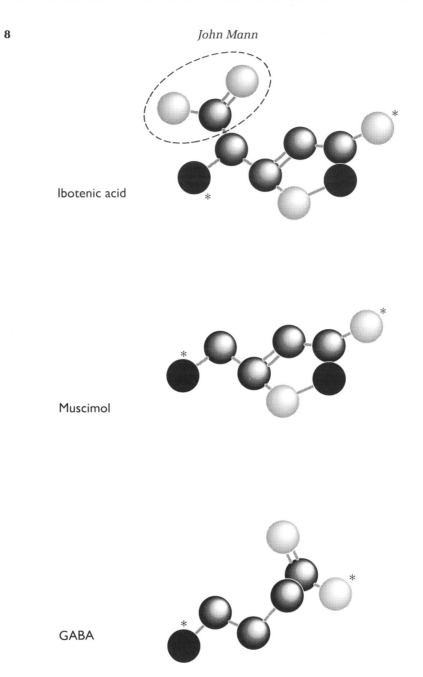

Ibotenic acid

Muscimol

GABA

Fig. 4 Comparison of the structures of ibotenic acid, muscimol, and GABA.

tions. So whilst it is not possible to experience exactly what the witches felt as they 'flew' to the Sabbat, it is interesting to compare the use of the plant-derived natural product as an adjunct to magic and clinical medicine.

The search for the elusive magic bullet

The Chinese Shennung Herbal of around 200 BC listed 365 plant- and animal-based remedies, while the Ancient Egyptians possessed the Ebers papyrus (*ca.* 1500 BC) which contained recipes for 800 potions that incorporated parts of crocodiles and hippopotamuses as well as more efficacious ingredients like opium and willow bark. Extracts of willow have, in fact, been used continuously for at least the last 3500 years for the treatment of inflammation and fever. The Ebers papyrus stated: *'When you examine a man with an irregular wound … and that wound is inflamed … you must make cooling substances for him to draw the heat out … leaves of willow.'*

In 1870, Marcellus von Nencki of Basel showed that extracts of willow bark could be used for the relief of fever in patients suffering from typhus and rheumatic fever, and went on to show that the major constituent, salicin, could be isolated and then converted (by hydrolysis) into salicylic acid, which was also efficacious. The first true clinical trial of this latter substance was carried out by the Scottish physician Thomas MacLagan in 1876. He treated more than 100 patients suffering from rheumatic fever and in each case complete remission of the symptoms—fever and swelling of the joints—was achieved. This was little short of miraculous given the usual severity and debilitation caused by this condition.

Unfortunately, both salicin and salicylic acid were very corrosive in the gastrointestinal tract. In an attempt to overcome this effect, the Bayer company in Germany prepared acetylsalicylic acid—or aspirin (Fig. 6)—and this was first marketed in 1899. It possessed the right mix of good anti-inflammatory and analgesic activity with a relatively low corrosivity; the rest, as they say, is history. Aspirin in all its various forms is still the most widely used drug for the treatment of pain and inflammation, and with the recent discovery that it helps to prevent coronaries in patients with arteriosclerosis, its future clinical utility is assured.

The mechanism of action of aspirin was not elucidated until 1970 when John Vane and coworkers showed that it inhibited the enzyme cyclooxygenase which controls the early stages in the production of the prostaglandins. These compounds are produced by many cells and help to control the normal functioning of the lungs, the gut, and the blood vessels, but are also involved in the production of inflammation and the associated pain and fever. Taken in normal doses, aspirin reduces the levels of prostaglandins and thus alleviates the symptoms associated with inflammation. More recently it has been demonstrated that in those with arteriosclerosis there are elevated levels of a particular class of cyclooxygenase, (called COX-2 to differentiate it from the

Fig. 5 An eighteenth century engraving of a witch being prepared for the Sabbat. Note the administration of the salve. (Wellcome Institute Library, London.)

normal enzyme, COX-1, present in healthy individuals). This new enzyme is particularly implicated in the production of a class of prostanoids that cause blood platelets to aggregate and is also sensitive to inhibition by aspirin, hence the utility of low-dose aspirin in the treatment of heart disease. Aspirin is thus an excellent example of a 'magic bullet' that can trace its lineage as far back as the folk medicine of Ancient Egypt.

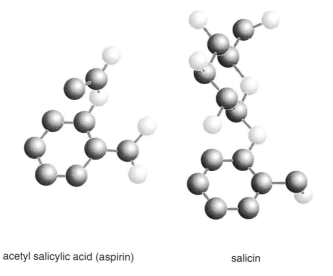

acetyl salicylic acid (aspirin)   salicin

Fig. 6 The structures of aspirin (acetylsalicylic acid) and salicilin.

Another drug with a well-defined pedigree is Ventolin or salbutamol, the best-selling bronchodilator used for asthma chemotherapy. The ancient Chinese used extracts of the plant *Ephedra sinaica* for the treatment of bronchial conditions including asthma and bronchitis and, when the major active constituent of the plant, ephedrine, was isolated in the 1920s, it was not too surprising when it was shown to be a potent bronchodilator. Unfortunately, its close structural similarity (see Fig. 7) to adrenaline—the 'fear, fight, and flight' hormone produced by the adrenals—ensures that ephedrine shares the potent cardiac stimulant activity of this latter substance. These effects are undesirable in an asthma drug and the pharmaceutical industry concentrated its efforts on the modification of the chemical structure and pharmacology of ephedrine. Salbutamol remains the optimum product of these efforts with its mixture of excellent bronchodilator activity combined with low cardiac stimulant potency; the structures shown in Fig. 7 demonstrate clearly the evolution of salbutamol from the ephedrine of Chinese folk medicine.

I would not want to give the impression that plants have provided all of the natural products that have entered clinical use: the importance of microorganisms as a source of drugs should be emphasized too. It is difficult to trace the lineage of penicillin back to some earlier folk use, but the practice of using mould extracts for the treatment of gunshot wounds was documented in Elizabethan times. In Norfolk it was usual to allow part of the Easter cake to grow mould and this was harvested 40

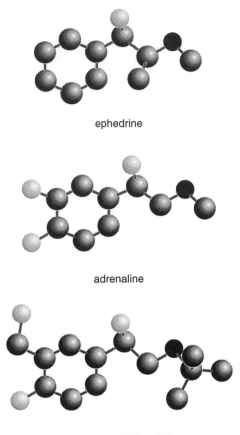

ephedrine

adrenaline

salbutamol (Ventolin)

Fig. 7 Similarity in structure between ephedrine and adrenaline.

days later, around Whit Sunday, and then stored in jars as a treatment for wounds and inflammation. Doubtless some of the moulds produced antibacterial substances, so Fleming's accidental discovery of the penicillins was in some ways foreshadowed by this deliberate cultivation of mould metabolites.

The antibiotic era dates from 1935 when the totally synthetic sulfonamides were introduced, and it is only in the years since 1945 that the antibacterial mould metabolites like the penicillins, cephalosporins, tetracyclines, and a plethora of even more exotic natural products have been isolated, identified, and commercialized. These new drugs have had an enormous impact upon the quality of life in this century. It should be recalled that in Victorian times average life expectancy was only around 45 years and that childhood mortality (from birth to age five years) was 150 in 1000. These premature deaths were in large measure attributable to the effects of bacterial infections, often as a sequel to viral infections like measles, mumps, influenza, etc., and the prevention of this mortality is a major triumph of modern chemotherapy.

The isolation of the penicillins from *Penicillium* species encouraged investigators to study the natural compounds produced by other microorganisms. This activity has continued to the present day and with the advent of high-throughout screening protocols during the past few years, these efforts have continued to reveal novel drugs. Many of these are antibiotics with novel or specific modes of action but others have other useful functions. Thus the cyclosporins from *Trichoderma inflatum* were discovered in 1976 and shown to possess marked activity as immunosuppressive agents. They have revolutionized the drug therapy associated with organ transplants. In the same year compactin was isolated from *Penicillium brevicompactum* and, together with the structurally similar mevinolin from *Aspergillus terreus*, was shown to be an excellent inhibitor of cholesterol biosynthesis. These two natural products have proved to be extremely effective in the treatment of life-threatening hypercholesterolaemia. The more recently discovered (1992) zaragozic acids from *Sporommiella intermedia* also inhibit cholesterol biosynthesis but by a different mechanism, and promise to be lead compounds for the design of new drugs which reduce the production of cholesterol. A plethora of microbe-derived anticancer agents have also been identified over the years, including the bleomycins from *Streptomyces verticillus*, the mitomycins from *Streptomyces verticillus*, and the anthracyclines from *Streptomyces peucetius*. These 'magic bullets' from terrestrial microorganisms will surely be followed by other discoveries once the massive biodiversity of marine organisms is further probed.

My three other topics concern diseases that are of worldwide importance: malaria, AIDS, and cancer. The first of these is generally viewed as a disease of developing countries—and so it is, with an estimated 300 million cases each year in Africa, South America, and south-east Asia, and around two million deaths, half of these children under the age of five. Until recent times, malaria was also a European disease, though the Jesuits provided some relief in the seventeenth century with their introduction (from South America) of the bark of the cinchona tree, the active ingredient of which is quinine. Malaria still represents a problem for those Europeans who travel to exotic places on holiday, because the parasite that causes malaria, *Plasmodium falciparum* in particular, has become resistant to most of the antimalarial drugs presently available. Each year several thousand European holiday travellers return with malaria and a few hundred subsequently die despite all the best treatment.

Once again the Chinese had a partial answer to the problem, for they have been using an extract of the plant *Artemisia annua* for at least the last 2000 years for the treatment of all kinds of fever including that associated with malaria. In 1971, the active component, artemisinin, was isolated for the first time and its structure elucidated. It was then used in a clinical trial during which 2099 malaria patients were treated including a number with cerebral malaria—the most dangerous type. Within a fortnight all had been cured.

This remarkable drug has at its core a peroxide (—O–O— linkage), a relatively rare chemical entity amongst natural products. Within the red blood cells that have been infected with the parasite (at the merozoite stage of its life cycle), this peroxide is cleaved to yield highly reactive radicals that bind to haemoglobin (thus denying it to the parasite), and also damage the parasite enzymes. As yet the parasite has no defence against this new 'magic bullet' with its novel mode of action, and the pharmaceutical industry both in the USA and in China is desperately seeking structural analogues of artemisinin with even more potent activity.

Our own modest endeavours in this area have centred around a number of ozonides that we have prepared which had sufficient structural similarity (see Fig. 8) with artemisinin to encourage us to have them tested for antimalarial activity. Gratifyingly, some of them are quite potent with activity at the microgram level against a resistant strain of *Plasmodium falciparum* from Thailand. This level of activity is about 1000 times lower than that observed with artemisinin, but our compounds can be prepared in three or four chemical steps, so could in principle be prepared on a large scale. At the moment we are trying to synthesize new ozonides with slightly more complex structures that are

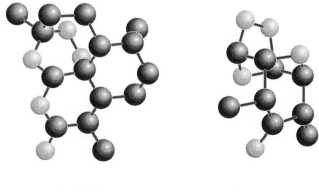

artemisinin   Reading ozonide

Fig. 8 Comparison of artemesin with a synthetic ozonide prepared at Reading University.

more closely similar to artemisinin, in the hope that these will have greater antimalarial potency.

Despite its widespread occurrence and devastating effects, malaria remains something of an orphan disease with little interest shown by the pharmaceutical industry in the development of new drugs. This is probably due to its relative unimportance in developed countries. AIDS, in contrast, receives a disproportionate amount of interest and research funding because it has a major impact on both developed and developing countries. The World Health Organization estimates that around 15–20 million persons worldwide are presently infected with HIV (human immunodeficiency virus), the virus that leads more or less inexorably on to the condition we call AIDS.

As can be seen in Fig. 9 there are three main stages in the life cycle of the virus at which drug intervention may have an effect. A number of substances, including the anticoagulant drug heparin, can associate with the surface glycoprotein (gp 120) of the virus and prevent its attachment

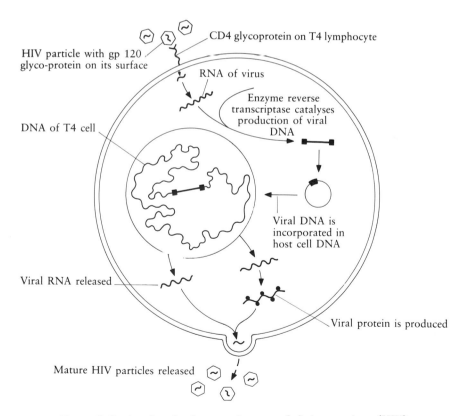

Fig. 9 Infection by the human immunodeficiency virus (HIV), and its replication within the cell.

to the surface glycoprotein of the human T4 lymphocyte. Without this association the virus cannot be internalized into the cell. Once inside the cell, the virus must use both its own enzymes and those of the host cell in order to make copies of its DNA and have these incorporated into the host cell DNA. One enzyme, viral reverse transcriptase, has been the target of most of the chemotherapy thus far. This enzyme catalyses the production of viral DNA using viral RNA as a blueprint (hence the term '*reverse transcriptase*', because in normal cells it is DNA that provides the blueprint for RNA production), and drugs like AZT, ddI, and ddC are effective inhibitors of this enzyme. While they cannot cure the disease they can at least slow the progression from the first infection to full-blown AIDS.

Our contribution in this area has centred on the synthesis of a new class of nucleosides which have a hydroxymethyl group at the position where AZT has an azide group. These are easily made from the carbo-hydrate D-mannitol using a novel photochemical reaction and, like AZT, they appear to inhibit the action of viral reverse transcriptase. An examination of the mechanism by which one nucleotide (the phosphate version of the nucleosides) is joined to the next during DNA biosynthesis (Fig. 10) reveals that they are connected between the 3'- and 5'-positions, and both AZT and our compounds should act as chain terminators. Initial biological evaluation by Wellcome showed that two of our compounds were about as potent as AZT and we have thus been encouraged to continue with this research.

Finally, maturation of new virus particles requires the attentions of a number of glycosidase enzymes which trim the newly formed viral coat glycoprotein. Several natural products, including castanospermine (from the Australian plant *Castanospermum australe*) and deoxynojirimycin (from various *Bacillus* species) inhibit these glycosidases, presumably because they resemble the natural substrates like glucose (see Fig. 11), and can bind to the enzymes, thus denying access to the glycoproteins that need to be trimmed. We are presently trying to prepare analogues of castanospermine which we hope will be more effective than the natural product.

While both malaria and AIDS are to some extent preventable or avoid-able, cancer is a disease that causes the most fear in developed countries due to its apparent indiscrimation and resistance to eradication. One in three persons in Britain contract cancer at some stage of their lives, but only one in five persons die from the disease. The difference between these two statistics reflects the major advances that have occurred during the past 25 years in surgery, radiotherapy, and, most importantly, chemotherapy.

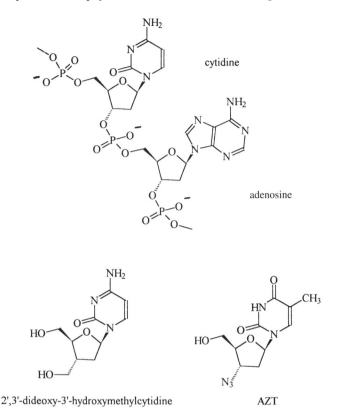

cytidine

adenosine

2',3'-dideoxy-3'-hydroxymethylcytidine

AZT

Fig. 10 The structure of a cytidine–adenosine dinucleotide, compared with the structures of two potential chain-terminators, 2',3'-dideoxy-3'-hydroxymethylcytidine and AZT.

The modern strategy of cancer treatment is illustrated in Fig. 12. At the time of diagnosis a tumour may well comprise a thousand million (10^9) cells and weigh about one gram. If it is localized, surgery and radio-therapy usually can be used to effect complete eradication. But all too often the primary tumour will have metastasized by the time of diagnosis and chemotherapy must be used in order to reach these disseminated sites of disease. If the number of tumour cells can be reduced to around ten thousand (10^4), the body's own immune system (if not too badly compromised) can then eradicate the residual disease.

Prior to the late 1960s there were a number of relatively unselective anticancer drugs available, most notably the nitrogen mustards that had evolved from the sulfur mustards used as blistering agents during World War I. These were not particularly effective in treating cancer, and their side-effects were severe. Once again the major advances in this area came from a study of folk medicine, and the story of the development

John Mann

Castanospermine

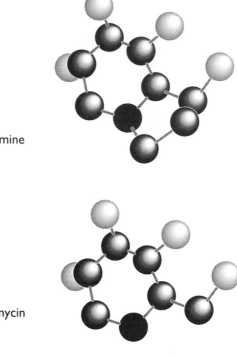

Deoxynojirimycin

Glucose

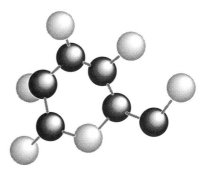

Fig. 11 Similarity between the structures of castanospermine, deoxynojirimycin, and glucose.

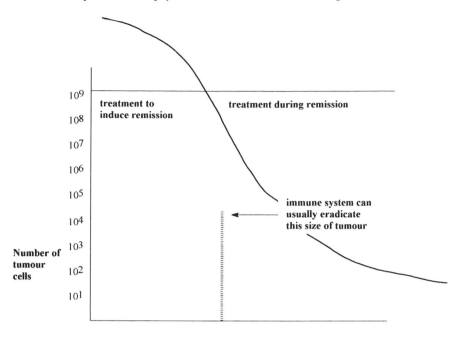

typicals weights of tumours: 10^9 cells - 1 gram

10^4 cells - 10 microgram

Fig. 12 If the size of a tumour can be reduced sufficiently, for example by chemotherapy, the immune system can usually then eradicate the tumour.

of the broad spectrum anticancer drugs vinblastine and vincristine provides an excellent example of this route of discovery.

Extracts of the plant *Catharanthus roseus*, the rosy periwinkle, had been used for centuries as a treatment for diabetes in the Philippines, South Africa, India, and other parts of south-east Asia. This encouraged R.L. Noble and his Canadian coworkers to screen these extracts for their hypoglycaemic activity, but they quickly established that it had no effect on rabbits whose glucose levels had been artificially increased. Some of these test animals subsequently died from severe bacterial infections and this was shown to be due to their inability to mount an efficient immune response. In particular, there had been a severe depletion of their white blood cells. Since this was a common side-effect of treatment with the existing anticancer drugs, the Canadian group went on to test their extracts against a variety of animal tumours, with excellent results. In 1955 they reported their findings and were immediately approached by scientists from the American pharmaceutical company Eli Lilley who had also been working with *Catharanthus roseus* and had made similar discoveries. After further collaborative research the groups were able to isolate and identify the major active alkaloids: vinblastine, vincristine, leurosidine, and leurosine; Lilley eventually marketed vinblastine and vincristine (in 1967) as broad-spectrum anticancer agents.

Vinblastine and vincristine are still widely used as components of multidrug regimes for cancer chemotherapy and these 'drug cocktails' have been spectacularly successful in the treatment of a number of hitherto untreatable conditions. For example, the five-year survival rates for patients with Hodgkin's lymphoma have increased from a depresssing 5% in 1970 to close on 100% with a regime that includes vincristine, the antimetabolites methotrexate and 5-mercaptopurine, and the anti-inflammatory drug prednisolone. Comparable figures for testicular teratoma would be 5–10% and around 95% using alternating treatments with the inorganic drug cisplatin and a combination of vinblastine and the mould product bleomycin.

The mode of action of vinblastine and vincristine involves a disruption of the process of cell division whereby a parent cell produces two copies of itself. This is known as *mitosis*. During this process the two developing daughter cells are held together by fibres known as *microtubules* until they are ready to lead a discrete existence. An enzyme, tubulin polymerase, controls the production of these microtubules and the two alkaloids bind to the enzyme and inhibit its actions. As well as inhibiting the division of cancer cells, the drugs unfortunately disrupt the division of normal cells in the stomach and to some extent in the bone marrow, hence the side-effects of gastric damage and decreased white cell counts. Fortunately, with careful dosing regimes these side-effects can be minimized and a high degree of selectivity can usually be achieved.

The success of these natural products prompted the establishment of large screening programmes at the National Cancer Institute in the USA and other centres around the world. Between 1960 and 1982, when the initial programme was terminated, the NCI screened about 114,000 plant extracts and perhaps twice that number of extracts of moulds and bacteria for anti-cancer activity. Since 1982 the screening programme has been expanded to test for other types of activity including anti-HIV activity, and increasingly the almost unexplored bounty from the sea has been investigated. The two compounds shown in Fig. 13, bryostatin A from a marine bryozoan and curacin A from a marine cyanobacterium, are currently of considerable interest due to their potent anti-cancer activity. We ourselves are part-way through a synthesis of curacin A and hope to make both the natural product (from fragments A–C) and structural analogues.

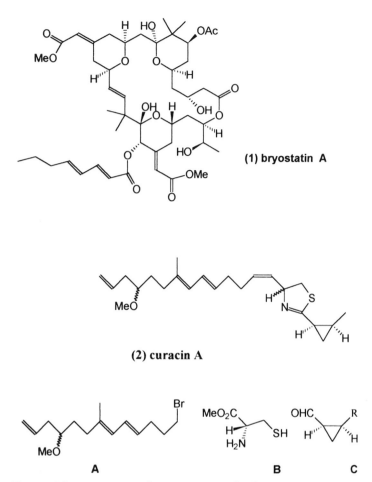

Fig. 13 The structures of two compounds showing strong anti-cancer activity: bryostatin A and curacin A.

Perhaps the most exciting new addition to the armoury of cancer chemotherapy is taxol, isolated from the Pacific yew, *Taxus brevifolia*. The yew has been used for centuries as a source of poisons, but it was not until 1971 that the chemical structure of taxol was determined and much more recently that its novel mode of anticancer activity was eluci-dated. It acts by disrupting the process of cell division, but in contrast to vinblastine and vincristine, taxol encourages the production of stable microtubules which do not easily break down. The two immature daughter cells are formed but cannot then break apart to become discrete new cells.

Unfortunately, taxol is not very abundant and it requires the sacrifice of a 100 year old tree to provide 3 kg of bark that ultimately provides 300 mg of pure taxol—about enough for one chemotherapeutic dose! Clearly this supply route is unacceptable on environmental grounds. The recent discovery that the foliage of the European yew, *Taxus baccata*, a renewable resource, contains a compound called 10-deacetylbaccatin III, which can relatively easily be converted into taxol, has greatly improved access to this important drug. It is presently undergoing clinical trials as the drug of choice for the treatment of advanced ovarian cancer and certain forms of lung cancer, two chemotherapeutic areas where there is a dearth of effective drugs.

Although a very large number of successful anticancer drugs are of natural origin, it would be wrong to give the impression that Nature has all of the answers. There are also a number of compounds that have been designed by chemists and pharmacologists to inhibit particular enzymes that are important for the growth of tumour cells; I should like to conclude by discussing some of our own work in this area.

Breast cancer is the most prevalent cancer amongst women and around 25 000 new cases are diagnosed each year in Britain. Of these tumours, between one-third and one-half are dependent upon a supply of the female hormones oestrogens if they are to grow, so one obvious strategy for the control of such tumours is to block the oestrogen receptors that occur in mammary cells. The Zeneca drug tamoxifen acts in this way and by preventing the association of oestrogens with their receptors, the subsequent activation of cell growth and division that would normally occur is disrupted. An alternative strategy is to inhibit the biosynthesis of oestrogens in the ovaries and this is the area in which we have made some contributions.

The drug Rogletimide or pyridoglutethimide was first made in the CRC laboratories at   Sutton and was shown to be an effective inhibitor of the enzyme aromatase, which catalyses the biosynthesis of oestrogens from the steroid androstendione (actually androst-4- en-3, 17-dione). This conversion involves the oxidative removal of one carbon atom (arrowed in Fig. 14) from androstendione and the enzyme uses an iron-containing cofactor (a haemin—rather like the cofactor used by the protein haemoglobin) to active molecular oxygen for this oxidation. The cofactor lies above the carbon atom that is to be removed and the activated oxygen species attacks from this top face. Molecular modelling shows that pyridoglutethimide can be superimposed on this lower half of the androstendione molecule (see Fig. 15) and it is easy to see how the nitrogen atom of the pendant (pyridine) ring (arrowed in the figure) might bind to the iron of the cofactor and thus deny access to oxygen. The result would be an inhibition of the conversion of androstendione into oestrogens and the pharmacological activity of the drug is thus explained.

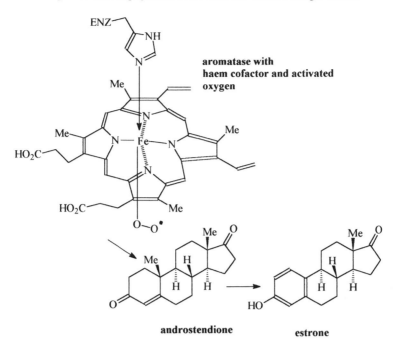

Fig. 14 Chemical species involved in estrogen biosynthesis.

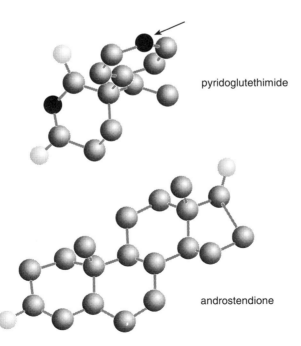

Fig. 15 Comparison of the structures of pyridoglutethemide and androstenedione.

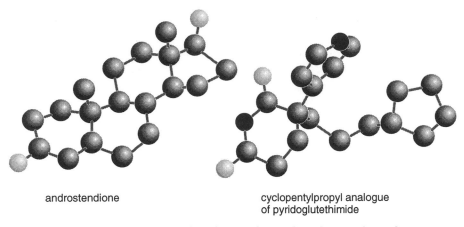

androstendione

cyclopentylpropyl analogue
of pyridoglutethimide

Fig. 16 The structure of cyclopentylpropyl analogue of pyrido-glytethimide, compared with androstenedione.

We became involved in this project when we were asked to devise a commercially viable (i.e. large-scale) synthesis of the drug and this we were able to accomplish. Our route has since been used to prepare multi-kilogram batches of pyridoglutethimide for the various clinical trials that have been conducted. One of the problems with Rogletimide is that it is not particularly potent and the patients need to take quite large doses of the drug. Our computer modelling suggested that the analogue shown superimposed on androstendione in Fig. 16 might fit into the active site of the aromatase enzyme even more effectively than Rogletimide and so we carried out the necessary synthesis. Biological evaluation of this new analogue was carried out at the CRC laboratories in Sutton and to our delight the compound was about one hundred time more potent (weight for weight) than Rogletimide as an aromatase inhibitor. This provides a nice example of how computer modelling, albeit rather basic in this instance, can be used to suggest structures that should be prepared for biological evaluation.

Our collaboration with the scientists at Sutton also led us into steroid chemistry and over the past dozen or so years we have prepared a large number of compounds that have good inhibitory activity against aromatase. One serendipitous discovery arose when the CRC group evaluated one of our compounds—4-fluoroandrostendione—against a different enzyme, so-called 5-alpha-reductase. This controls the metabolism of testosterone, the main male hormone, to produce dihydrotestosterone. Disturbances in the balance between these two steroids are implicated in such diverse medical conditions as acne, male-pattern (premature) baldness, and, most importantly, the onset of prostatic hyperplagia (overgrowth of the prostate) and prostatic cancer. Drugs for the control of this enzyme are clearly of enormous importance and so we were excited to learn that 4-fluoroandrostendione was a very effective inhibitor of the enzyme. Unfortunately, it possessed some activity as a male hormone which was undesirable, so we prepared a number of analogues based upon the female hormone progesterone and one of these, 4-cyanoprogesterone, was a very potent inhibitor indeed. This has provided the incentive for more research and we are presently trying to prepare other steroids with an even better biological activity profile.

I have wandered away from my main theme of the utility of plants as a source of drugs, so let me conclude by re-emphasizing the importance of this resource. It is estimated that there are about 250 000 terrestrial higher plant species with around 10 000 still to be identified. Of these species only about 15 per cent have been systematically evaluated for biological activity. Since many of these grow in the tropical forests, which are rapidly dwindling in size due to Man's activities, many potential drugs may never be identified, let alone developed. If we turn to the oceans, which cover 70 per cent of the planet's surface and probably contain about half of the total global diversity in terms of life-forms, we are confronted with an almost infinite number of unidentified and unexploited organisms. Estimates of total biodiversity range from around 3 million to an upper limit of 500 million species. Most of these are microorganisms rather than land or sea creatures and plants, but one has only to recall the importance of Fleming's discover of the penicillins or the more recent discovery of cyclosporin, the inhibitor of organ transplant rejection, to appreciate how important these microorganisms might become.

Natural products from plants, animals, and microorganisms still provide the source of around 50 per cent of our modern drugs, and higher plants are responsible for about half of this total. Our debt to Nature and to the primitive societies who first used these plants is thus obvious. There is still much to learn. The following words of Paracelsus, the sixteenth century apothecary and alchemist, and those of William Shakespeare are as apposite today as they were when first written.

> *All substances are poisons; there is none which is not a poison;*
> *the right dose differentiates a poison and a remedy.*
> (Paracelsus)

> *O! mickle is the powerful grace that lies,*
> *In herbs, stones, and their true qualities:*
> *For nought so vile that on earth doth live*
> *But to earth some special good doth give,*

Within the infant rind of this weak flower
Poison hath residence and medicine power.
                    William Shakespeare (*Romeo and Juliet*, II, iii)

Further reading

General background
J. Mann, *Murder, Magic, and Medicine*, Oxford University Press, 1992.

Absinthe
W.N. Arnold, *Sci. Amer.*, 1989 (June), **258**, 86.

Hallucinogenic plants
A. Hofmann and R.E. Schultes, *Plants of the Gods*, Van der Mark Editions, 1979
    New York.

Aspirin
G. Weissman, *Sci. Amer.*, 1991 (June), **264**, 58.

Antibiotics
R.G. Macfarlane, *Alexander Fleming: The Man and the Myth*, Oxford University
    Press, 1985 Oxford.
B. Dixon, *Power Unseen—How Microbes Rule the World*, W.H. Freeman, 1994
    Oxford.

Malaria
P. Brown, *New Sci.*, 1992, 31 October, 37.
L.J. Bruce-Chwatt and J. de Zuluetta, *The Rise and Fall of Malaria in Europe*,
    Oxford University Press, 1980 Oxford.

AIDS
P. Brown, *New Sci.*, 1992, 18 July, 31.
R. Weiss, *Science*, 1993, **260**, 1273.

Cancer
G.A. Cordell, *Chem. and Ind.*, 1993, 1 November, 841.
W.B. Pratt and R.W. Ruddon, *The Anticancer Drugs*, Oxford University Press,
    1995 New York.

JOHN MANN

Born 1945, educated at Dartford Grammar School and University College,
London. After postdoctoral research in the USA (Syntex and Harvard)
and at Oxford, he moved to a Lectureship in Organic Chemistry at Reading University in 1974, and was subsequently promoted to a Personal
Chair in Organic Chemistry in 1990. His research interests include the

isolation and synthesis of natural products and related structures, and the rational design and synthesis of drugs for the treatment of cancer, AIDS, and Alzheimer's disease. He has published around 100 research papers as well as five text books and a popular science text, *Murder, Magic, and Medicine* (published by Oxford University Press), upon which this talk is based.

The Woodhull Lecture: visual art and the visual brain

SEMIR ZEKI

Introduction

There are certain truths about art and the visual brain that are so self-evident that we may accept them as being largely axiomatic. Chief among these is that we do not see with our eyes but with our brains, the eyes being nothing more than an essential filter and conduit of visual signals to what is known as the primary visual cortex, area V1 (Fig. 1), of the brain which itself redistributes the visual signals to yet further visual areas which we have discovered only in the past twenty years or so. No visual art is therefore possible without the visual brain, from which it follows that all art, whether in conception, in execution, or in appreciation, must obey the laws of the brain.

The law of functional specialization

The first of these laws is the law of functional specialization[1], by which I mean that different attributes of the visual scene, such as form, colour, and motion, are processed in geographically separate parts of the visual brain. Experimental evidence has shown that cells in the visual brain are remarkably selective; first they respond to stimulation of only a small part of the field of view, known as their receptive fields, and secondly they are selectively responsive to only some kinds of stimuli falling in their receptive fields. Some, for example, respond specifically to light of certain wavelengths and not to other wavelengths or to white light, others respond to lines of specific orientation and not to other orientations, and yet others respond to motion in one direction and not to motion in other directions (Fig. 2). Cells that are selective for a certain attribute are

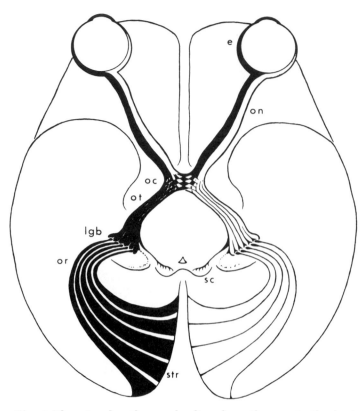

Fig. 1 The visual pathways leading from the eye to the brain. e, eye; on, optic nerve; oc, optic chiasm; ot, optic tract; lgb, lateral geniculate body; or, optic radiation; sc, superior colliculus; str, striate cortex.

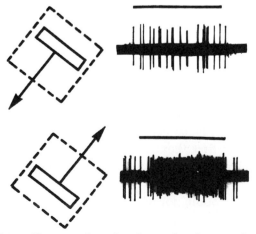

Fig. 2 This cell responds only when a bar is moved within its receptive field (dotted outline), and further, it responds when that motion is in one direction only (towards 1 o'clock).

grouped together in particular areas of the cerebral cortex, thus conferring their specialization on the areas.

A simple way of demonstrating this in the human brain is to let humans look at particular types of visual stimuli, in which particular attributes such as motion or colour are emphasized, and measure the change in the cerebral blood flow, itself indicative of an increase in activity in specific parts of the brain (Fig. 3). When one does so, and examines horizontal slices through the brain in which areas of highest activity are shown in white, one finds that looking at an abstract multicoloured scene with no recognizable objects activates not only area V1, through which visual signals enter the brain, but another area, the 'colour centre', located in the

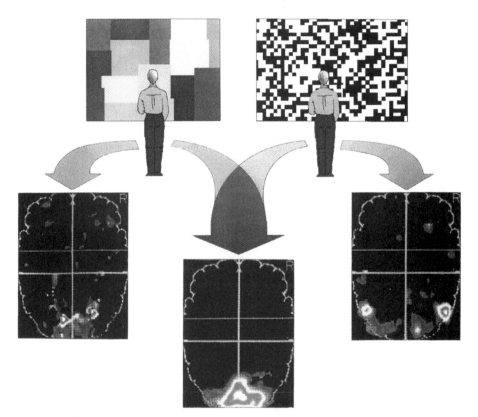

Fig. 3 A subject is placed in a Positron Emission Tomography (PET) scanner, which detects increases in cerebral blood flow. Whilst in the scanner the subject is shown a static multi-coloured display, and increased activity is detected in an area known as V4 (left). If the subject is shown a pattern of moving dots, activity is detected in another area, known as human V5 (right). In both these conditions, areas V1 and V2 are active (centre). (After Zeki S., *La Recherche*, 1990; **21**, 712–721).

cortex surrounding area V1, to which we refer as area V4. On the other hand, when one stimulates with a pattern of small black and white squares that move in different directions, one finds that this time, in addition to the activity in V1, there is another area in the visual brain outside V1, quite separate from the colour centre, which becomes active, an area which we refer to as the 'motion centre' or area V5[2].

A second law follows from this demonstration, namely that the brain assigns special areas to the processing of those visual features that are of particular significance to the organism. Colour, form, motion, faces, facial expressions, and even body language fall into this category and all have their separate representational seats in the cortex. I put forward the seemingly obvious proposition that it is those attributes that are separately mapped in the visual brain that have primacy in art. I do not mean to imply that the aesthetic effects of colour depend uniquely upon the area of the visual brain which is specialized for colour but only that that area is necessary for any colour experience at all and that without it no aesthetic experience of colour is possible. Consider this: when the colour centre is damaged as a consequence of a stroke, the result is cerebral achromatopsia or an inability to see the world in colour[3]. Patients with such a lesion uniformly describe the world as consisting of dirty shades of grey. They are no longer able to obtain one particular type of knowledge about the world. Colour perception is not possible for an achromatopsic patient and he can therefore have no aesthetic experience in that domain. It is little use asking such a patient to admire the subtleties of fauvist art or to appreciate the mature *poésie* of Titian, expressed in colour. Indeed, one of the achromatopsic patients that I examined, the patient of Oliver Sacks, was himself an artist, and a great admirer of Impressionist art as well as the art of Vermeer. He told me that since becoming achromatopsic, he could no longer bear to go to an art gallery.

Or consider portrait painting, which has been a dominant feature of Western art, as another example. One function of portrait painting, especially in the days before photography, was to acquaint or to remind men and women of what their loved ones or future spouses look like, because the recognition of individuals is most often and most easily done through their faces. It is because of this that the brain has devoted an entire area to the recognition of faces, which in turn explains why portrait painting has been such a dominant feature in our art.

Perhaps the first thing to notice about a portrait is that the face usually dominates the painting even if it does not constitute the predominant part in terms of size or luminosity. There are many examples of this; especially prominent among these are many of Rembrandt's portraits as

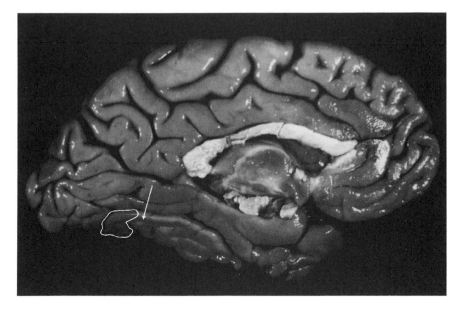

Fig. 4 The medial surface of the human brain. The centre in the fusiform gyrus has been highlighted. Damage to the arrowed area of this gyrus leads to the condition of prosopagnosia, or the inability to recognize familiar faces.

well as those of his contemporaries, where the intensity of light reflected from the then fashionable collars is much greater than that reflected from the face, which is in fact sometimes half obscured. Apparently, the brain is much more interested in focusing and concentrating on the face—because it yields a good deal more information. The rest of the painting is a sort of prop, enhancing aesthetically the portrait but not necessary to it—the face can survive on its own. The area of the brain that is critical for the recognition of familiar faces lies ventrally in the occipital lobe, close to the colour centre (Fig. 4, arrowed region), and damage to it leads to the condition of prosopagnosia, or an inability to recognize familiar faces. In this condition, the brain is no longer able to obtain visual knowledge about a particular person through the face. It is in many ways an extraordinary syndrome, when the subject can recognize the details of a face, the eyes, the nose, the mouth, and so on but cannot put all these details together to recognize a particular face[4]. There is one extraordinary and somewhat frightening description of a patient who suffered a stroke that struck the relevant area while he was with his physiotherapist. 'What is happening, Mademoiselle,' he said, 'is that I can no longer recognize your face', even though he knew perfectly well who she was[5]. Such patients often have to use other features, such as the voice, to recognize familiar faces. Again, I do not maintain that the aesthetic quality of a portrait is uniquely dependent upon the relevant area in the fusiform gyrus, but only that the recognition of familiar faces, and therefore one of the aims of portrait painting, is not possible without this area, from which it follows that there can be no aesthetic quality associated with a particular face when a portrait is viewed by a prosopagnosic patient.

People of course change and die and memories of what they looked like soon fade away. The fact that the art of portrait painting outlives the person portrayed derives from the fact that portraits tell us a good deal more than what a person looked like, and a glance at a portrait may tell us whether the person portrayed is sad or happy, arrogant or humble, and much else besides. Portraits are therefore a way of giving the spectator knowledge about the personality of someone, even someone whom we have never met. Titian's portrait at the National Gallery, said to be of himself, is that of a man recognizable at a glance as being somewhat arrogant and disdainful. To that extent, it is representative of a host of such faces and therefore of such people and to that extent there is a constancy about it since it captures the essential ingredients of what the brain interprets as that type. Artists have developed a great many tricks to communicate to us, visually and through our brains, the characteristics that mark a person. In his portrait Titian uses the device of the twisted view, apparently then common in Italy[6], to enhance the effects of self-assuredness, his subject looking at us with his eyes only, his head being only partially turned in our direction. There are many other examples one could give but perhaps the above suffices to make the point that small and subtle changes, especially in the eyes, can make a big difference to the brain's perception of faces and its ability to acquire knowledge about them.

The fact that my brain as well as yours can categorize at a glance the Titian portrait as that of a haughty and self-confident person effectively means that Titian (or his brain) managed to capture on canvas an essential feature which gives immediate knowledge about that person. Whether the portrait itself bears any likeness to Titian or not is immaterial, except perhaps to Titian himself and to those who knew him well. Perhaps Michelangelo was right when, reproved because his sculptures for the Medici tombs in Florence bore little resemblance to the Medicis buried inside, he replied, 'In a thousand years, who will remember what the Medicis looked like'.

Now it is, to me at any rate, an astonishing fact about the organization of our visual brain that it is capable of separating recognition of a face from the expression on a face[7]. If you take a prosopagnosic patient, one who can no longer recognize familiar faces, and ask him to describe whether the expression on a familiar face that he is no longer able to recognize, is that of a sad or happy person, he can do so with accuracy. It is only when lesions extend more anteriorly in the fusiform gyrus that the perception of a face and of the expression on it are abolished. One is led to the perhaps surprising conclusion that, because of the important knowledge that both the face and the expression on it can give, separately, the brain has allocated separate areas for each.

Functional specialization in visual aesthetics

We can now ask whether, in neurological terms, there is a single unified sense of aesthetics or whether there are many different aesthetics, each one linked to a separate system. Whatever the philosophical or artistic answer to that question, there seems little neurological doubt that there are many different aesthetics within the domain of vision, attached to say colour, or to expressions, or to movements, or to forms, that each is tied to a separate set of cortical areas, and that the destruction of one need not entail the destruction of all. I am tempted to generalize and say that there is a functional specialization in aesthetics, which does not exclude the fact that all of these separate aesthetic systems may both interact with one another and also lead to a higher aesthetic which neurology has not even begun to investigate. In the same way, the fact of functional specialization does not imply that the different systems do not interact with each other and lead to the final image in the brain, which is a synthesis of all the different attributes, undertaken by the brain in a way that we still do not understand.

The law of constancy

The third law is perhaps the most fundamental of all the laws regarding the visual brain and it is in it that I find a direct link between the functions of art and the functions of the brain. I will call it the law of constancy and both its significance and its relation to art will become clearer if one were to ask the most fundamental of all questions regarding the visual brain, surprisingly a question that is never asked in practice: why do we need to see at all? Different people would probably have different answers to that question; few, I imagine, would believe that we see in order to be able to appreciate art. Most would perhaps give answers such as: in order to be able to recognize people, or to find your way about, or to acquire food, or to read. Yet none of these answers is satisfactory, because none is broad enough. Many animals, among them mice and moles, have very rudimentary vision, if indeed they have any at all, and are yet fairly successful in negotiating their way about their environment and generally in undertaking such activities which have allowed them to survive successfully. The answer to our question is, I believe, much simpler and more profound—*we see in order to be able to acquire knowledge about this world*[8]. Vision is not of course the only sense through which we can acquire that knowledge. Other senses do just the same thing. Vision just happens to be the most efficient way of acquiring knowledge, which is perhaps one reason why so large a part of our brains, amounting to perhaps one-quarter of the total, is devoted to vision. Moreover, there are certain kinds of knowledge, such as the colour of a surface or the expression on a face, that can only be acquired through vision.

Such a definition of vision is not one voiced by neurologists and I have never encountered it among artists, though I may of course be ignorant of much that they have said. Yet it is perhaps the only definition that unites neurology and art, that finds a common thread linking the functioning of the visual brain to the function of visual art, which is itself one of the products of the brain. It is, at any rate, a definition worth exploring, because in it I find a general and unifying theory of visual aesthetics, one that encompasses the views of Plato and Michelangelo, of Schopenhauer and Heidegger, no less than those of Cézanne, Braque, and Mondrian, but one which, unlike theirs, is based on the functions and functioning of the visual brain.

It requires little to realize that the acquisition of knowledge by the visual brain cannot be an easy matter. Our visual environment is in a continual state of flux and objects and surfaces are viewed from different angles and distances, and in different lighting conditions. But the brain is only interested in the permanent properties of objects and surfaces. Vision is, therefore, a search for essentials. To extract those essentials from the ever-changing information reaching it, the brain must undertake three interdependent processes: it must discount and discard much of the information that reaches it, it must select from that information only that which is necessary for acquiring knowledge about the permanent, constant, and essential properties of objects, and, finally, it must compare the selected information with its stored record and hence identify and categorize an object as belonging to one or another group of objects. One can see that the first law, of functional specialization, is intimately connected to the third, the law of constancy, because the kind of information that the brain has to discard to get at one attribute of the visual scene, say its colour, is very greatly different from the kind of information that it has to discard to get at another attribute, say the size; in the former, the brain has to discount the illuminant in which a colour is viewed and in the latter the viewing distance. Hence the brain has dedicated special areas for extracting the constants related to different attributes.

Vision is not, therefore, the passive process that we have for so long supposed it to be, but an active one in which the brain constructs the visual image and that final visual image depends as much upon the external physical reality as it does upon the operations of the brain. Consider colour: the brain is able to assign the colour 'green' to a leaf whether the leaf is viewed at noon on a cloudy or sunny day or at dawn or dusk. If one were to measure the wavelength composition of the light reflected from the green surface objectively one would find considerable variations, including at times an excess of long-wave (red) light—it is as if the apparent, objectively measured, 'truth' tells us that the surface is not always green. Yet, by comparing the wavelength composition of the light reflected from the leaf with that which is reflected from the surround, the brain is somehow able to discount the objective measure and to assign a constant colour, green, to the surface. The ability of the brain to assign a constant colour to a surface, in spite of wide-ranging differences in the illumination conditions under which that surface is viewed, is one that Helmholtz[9] called the process of '*discounting the illuminant*' and thought that it is done by an unconscious inference by the brain in its desire to get to the essence of things; today we call it 'colour constancy' and know that it is the product of neural interactions within the colour centre. Gleizes and Metzinger, in their book on Cubism, wrote of the necessity for an artist to '*sacrifice a thousand apparent truths*' to get to the essence of things, a statement remarkably similar to that of Helmholtz[10]. A physiologist, in describing the functions and functioning of the visual brain, could hardly improve on that statement. The final visual image in the brain is thus dependent upon both the external, physical, reality and on the rules and activities of the brain.

This analysis of the functions of the visual brain not only illustrates why there is a law of constancy, but also provides us, I think, with the beginnings of what I shall call a general constancy theory of visual art. Such a theory would define art as an extension, or a manifestation, of the functions of the brain in the quest for constancies and therefore for essentials.

The neurology of the Platonic ideal

In Athens, some two and a half thousand years ago, the circle of Plato talked about these problems, and summarized the problem of painting as they saw it. Consider the following lines, taken from Book X of Plato's *Republic*:

> Does a couch differ from itself according as you view it from the side or the front or any other way? Or does it differ not at all in fact though it appears different, and so of other things?
>
> That is the way of it, he said. It appears other but differs not at all.
>
> Consider then this very point. To which is painting directed in every case, to the imitation of reality as it is or of appearance as it appears? Is it an imitation of a phantasm or of the truth?
>
> Of a phantasm, he said.
>
> Then the mimetic art is far removed from the truth.
>
> Yes, he said, the appearance of form, but not the reality and the truth

(from ref. 11)

To Plato, then, painting was a relatively low art, a mimetic art, one that could only represent one aspect of a particular example of a more general category of object. Indeed, given a chance he would have banished all painters from his millenial Republic since they could only capture one facet of the truth. To him, and to the ancient Greeks in general, there was the general ideal, the ideal couch in this instance, which was the embodiment of all couches; then there was a particular couch which was but one example of the more general, 'universal', couch; and, finally, there was painting, which captured but one facet, one image, of one particular couch. 'The Greeks', Sir Herbert Read[12] tells us, 'with more reason, regarded the ideal as the real, and representational art as merely an imitation of an imitation of the real'.

The example that Plato gives above, that of a couch, is an interesting one in that a couch is not necessarily associated in most minds with great beauty or aesthetic appeal. This choice is probably deliberate, for the view expressed in the passage is only one example of a more general theory of form. If we ask what is a couch, we do not ask about a particular couch but instead enquire into what all couches have in common, in other words we ask about that property which enables us to categorize them as couches. The common elements identify them. So what Plato was really saying was that a single view or image of a particular couch, depicted in a painting, could not be representative of all couches and could not therefore give knowledge of all couches. Without saying so explicitly, and almost certainly without realizing it, he was really comparing the 'phantasm' of painting with the reality of perception, a function of the brain, where there is no problem with a particular facet or view, because the brain usually has many views of the same object and is able to combine all the views to acquire knowledge about an object. Plato therefore implied that painting should strive to expand and possibly change direction in such a way that, by viewing one painting alone, we should be able to acquire knowledge about all objects of that category represented in the painting. What he only implied, Schopenhauer made explicit many centuries later, when he wrote that painting should strive 'to obtain knowledge of an object, not as a particular thing but as Platonic Ideal, that is the enduring form of this whole species of things'[13], a statement that a modern neurobiologist could easily accommodate in describing the functions of the visual brain. Painting, in other words, should be the representation of the constant elements, of the essentials that would give knowledge of all couches; it should, in brief, represent constancies. As John Constable put it in his *Discourses*[14]: '... the whole beauty and grandeur of Art consists ... in being able to get above all *singular forms, local customs, particularities of every kind* ... [The painter] makes out an abstract idea of their forms more perfect than any one original', the 'abstract idea' being presumably Constable's term for the Platonic Ideal.

It is not difficult to see that, in the opinion of Plato and other like-minded philosophers, painting could be described in neurobiological terms as a search for constancies, a means of getting above all, 'singular forms [and] particularities of every kind', in fact of achieving precisely what the brain does so effortlessly. The brain is interested in particularities, but only with the broader aim of categorizing a particularity into a more general scheme. For the brain, a couch is categorized immediately as something that you lie down on or sleep in, provided it is given a sufficient amount of information to identify it as such. This identification is dependent upon the brain's stored memory of couches in general, and is not therefore dependent upon any given view because the brain has already been exposed to many different views of many different couches; any one of these is sufficient to allow it to classify a couch as a couch. As Gertrude Stein might have said, for the brain, a couch is a couch is a couch, just as a rose is a rose is a rose.

In neurological terms, therefore, the Platonic Ideal is nothing more than the brain's stored representation of the essential features of all the couches that it has seen and from which, in its search for constancies, it has already selected those features that are common to all couches. We know a little, but not much, about the brain's stored visual memory system for objects. We know that it must involve the inferior convolution of the temporal lobes because damage here causes severe problems in

object recognition. Although very much in their infancy, recent physiological studies[15] have started to give us some insights into the more
detailed physiological mechanisms involved. When a monkey, an animal
that is closely related to humans, is exposed to different views of objects
that it has never encountered before (objects generated on a television
screen), one can record from single cells in the inferior temporal cortex
to learn how they respond when these same objects are shown on the
television screen again. Most cells discharge in response to one view
only, and their response declines as the object is rotated in such a way as
to present increasingly less familiar views. A minority of cells respond to
only two views but only a very small proportion, amounting to less than
1 per cent, respond in a view-invariant manner. Whether they respond to
one or more views, the actual size of the stimulus or the precise position
in the field of view in which they appear makes little difference to the responses of the cell. On the other hand, no cells have ever been found that
are responsive to views with which the animal has not been familiarized;
hence exposure to the stimulus is necessary, from which it follows that
the cells may be plastic enough to be 'tuned' to one or more views of an
object. In summary, many cells, each one responsive to one view only,
may be involved during recognition of an object, the whole group acting
as an ensemble.

Interesting though such cells are, they cannot present the entire physiological background to object recognition. We know that this is a property
that must be very widely distributed in the brain, a supposition that
follows directly from the functional specialization of the many, widely
distributed, visual areas. That it must be very widely distributed and
require the co-operation of several areas is also shown by the fact that,
except for lesions of V1, which lead to total blindness, there is no known
example of a lesion restricted to the cortex surrounding V1 which dis-
rupts recognition of all aspects of the visual world or indeed of all
shapes and objects. We also know that the cerebral mechanism for elicit-
ing different visual memories may in fact differ. We know, finally, that
the temporal lobe and structures in its vicinity, such as the hippocampus, are involved, partly because electrical stimulation of these
regions reawakens long-forgotten memories and partly because damage
to them, and especially the hippocampus, leads to severe problems of
memory. But of the detailed mechanisms we are more or less ignorant.

We can now begin to see that there is a straightforward relationship
between the Platonic Ideal and the brain-based concept of constancies. A
couch may be said to have certain constant features, no matter what
angle one views it from, and it is these constant features, the ones that it
shares with all couches, that are represented in the brain. Likewise the

Platonic Ideal of a couch is what is common to all couches; it is in fact the brain's stored record.

The aims of Cubism

It was Cubism that set out to address that deep paradox between reality and appearance in representational art and its absence in ordinary visual perception that Plato had alluded to. Cubism was inaugurated by Picasso and Braque as one of the most radical departures in Western art since Paolo Ucello and Piero della Francesca introduced perspective into painting. The precursor of this style is generally accepted to be Picasso's *Les Demoiselles d'Avignon* (Fig. 5) where the beginnings of the effort to eliminate both perspective and the point of view are evident. Especially notable is the figure to the bottom right, who could be sitting sideways, or facing us, or facing away from us and, to a lesser degree, there is also an ambiguity in her face. That ambiguity is taken to greater lengths in *Portrait of a Woman* (Fig. 6) and heightened further in later representative

Fig. 5 *Les Demoiselles d'Avignon* by Pablo Picasso. (The Museum of Modern Art, New York.)

paintings such as the *The Violin Player*, where Picasso approached the naturalistic image from so many different viewing angles and combined these views onto canvas (attempting in a way to represent the real violin player, not a particular violin player) that the final product is barely recognizable as a violinist, save through its title. Hence Picasso and Braque were trying to achieve a sort of constancy—the major attribute of the visual brain. And here you begin to see the limitations of art when compared with the infinite capacities of the brain. The brain can unite these separate views into one, and the result is not the unrecognizable image that Picasso's painting presents.

Fig. 6 *Portrait of a Woman* by Pablo Picasso.

The aims of Cubist painting were stated at the beginning of this century by Jacques Rivière[16]. Without referring to Plato, what he wrote is almost identical to what Plato had said some two thousand years before, and like Plato, made without reference to the brain, though it is doubtful if a neurologist could find more adequate words to describe the operations and functions of the brain:

> The Cubists are destined to give back to painting its true aims, which is to reproduce objects as they are. But, to achieve this, lighting must be eliminated because it is the sign of a particular instant. If, therefore, the plastic image is to reveal the essence and permanence of things, it must be free of lighting effects. It can therefore be said that lighting prevents things from *appearing as they are* ... Contrary to what is usually believed, sight is a successive sense; we have to combine many of its perceptions before we can know a single object well. But the painted image is fixed. Perspective must also be eliminated because it is as accidental a thing as lighting. It is the sign, not of a particular moment in time, but of a particular position in space. It indicates not the situation of objects but the situation of a spectator. Perspective is also the sign of an instant, of the instant when a certain man is at a certain point.

But that is precisely what the brain does—to reproduce objects as they are, no matter what angle they are viewed from.

Given this very close similarity between the aims of the visual brain and the aims of art, one may indeed regard the aim of the latter as being nothing less than an extension of the aim of the former. And given this similarity, it is surprising that the connection between the two has never been made before. The reason for this is simple. It is only in the past few years that we have begun to understand that vision is not a passive process but an active search for essentials. Previous to that, we commonly thought of vision as being nothing more than a simple photographic process, with an image of the visual world, in all its forms, colours, and movements, being impressed on the retina and then transmitted to be received, passively, by the visual receiving area, V1, in the cortex. The image thus received would be interpreted by the cortex surrounding V1, called the visuo-psychic cortex, thus leading to the understanding of what was seen. Vision was therefore conceived of as a dual process, of seeing and understanding, with each faculty having a separate cortical seat. That view has its basis in a simple and powerful anatomical fact, namely the nature of the connections between retina and cortex and the consequences of damage to these connections or to the visual receptive cortex. For it was, until very recently, one of the best established facts about the visual brain that the retina connects with V1 only, with adjacent points on the retina connecting to adjacent points on V1, thus recreating a map of the retina on the cortex (Figure 1). The consequence of damage to this pathway, or to the visual receptive cortex (V1), is blindness. The extent and position of the blindness is proportional to the extent and position of the damage: the blindness is total if the whole of V1 is damaged and sub-total if the damage is only partial. The cortex surrounding V1 was thought not to receive an input from the retina, but to receive its visual input from V1 instead. It is no wonder, therefore, that the founding fathers of our subject considered that seeing consists of a passive process, that it is with our V1 that we see and that it is with the vaguely defined cortex surrounding it that we understand what we see, without realizing that seeing is, in a sense, understanding and that the two processes are not easily separable.

Our new concept of the functions of the visual brain allows us better to consider art as being an extension of the functions of the brain in its search for essentials. And great art can thus be defined, in neurological terms, as that which comes closest to showing as many facets of the reality, rather than the appearance, as possible and thus to satisfying the brain in its quest for essentials.

A neurological excursion into the art of Vermeer and Michelangelo

With that thought in mind, it is perhaps interesting to consider Jan Vermeer's *Man and Woman at the Virginal*, now in Her Majesty's collection. I thought that I ought not to pass judgement on any painting. Who am I, after all, to pronounce on the quality of these works? But in the end the judgement cannot be avoided, although I do so with diffidence and humility and then only as a neurobiologist. The painting, I believe, derives its grandeur and unique position not only from its technical virtuosity but also from its ambiguity, by which I mean its ability to represent simultaneously, on the same canvas, not one but several truths, each one of which has equal validity with the others. These several truths revolve around the relationship between the man and the woman. She could be his sister, his pupil, his wife, or his lover; he could be just listening to her playing, or announcing a separation or a reconciliation. All these scenarios have equal validity in this painting which can thus satisfy several 'ideals' simultaneously—the brain can recognize this and categorize it as representative of a happy or sad event. This gives ambiguity—which is a characteristic of all great art—almost a different, and neurological, definition, not the vagueness or uncertainty found in the dictionaries, but on the contrary, certainty; the certainty of many different, and essential, conditions, each of which is equal to the others, all expressed in a single profound painting, profound because it is so faithfully representative of so much.

If Vermeer achieved his effect in a finished painting, another, who has a place among the highest, at times tried to achieve the same multi-representation in exactly the opposite way. It is well known that Michelangelo spent much time and effort trying to portray not only physical beauty but also spiritual beauty and divine love, particularly in relation to the Descent from the Cross. It is equally well known that he often left his sculptures unfinished. Perhaps the most famous among these are his two sculptures, the *San Matteo* and the *Rondanini Piéta*, but there are others. By deliberately leaving them unfinished, he allowed the viewer to become imaginatively involved and perhaps see many possible, equally valid, designs and forms, not one. As Gleizes and Metzinger were to say much later in their book *On Cubism*, 'Some forms must remain implicit, so that the mind of the spectator is the concrete place of their birth'.

To best appreciate Michelangelo's aims and approach, it is as well to look at his *Sonnets*, where he best expressed his theory of art and beauty. In one of them, *The Lover and the Sculptor*, dedicated to Vittoria Colonna, he tries to tell us that the artist experiments to discover what the brain can really see. He wrote:

> The best of artists have no thought to show
> That which the rough stone in its superfluous shell
> Does not include: to break the marble spell
> Is all that the hand that serves the brain can do.†

The pathology of aesthetics

But, Plotinus, another Neo-Platonist and one whose writings Michelangelo was surely acquainted with, had said many centuries before him, 'the form is in the designer long before it ever enters the stone'. It is, of course, in the viewer as well, which is precisely why forms can remain implicit. But if the form is in the designer before it ever enters the stone, then it is just possible that the artist needs no external scene to inspire his art. This is, in fact, the starting point of the non-objective art of the

---

†I have used the translation by Symonds; other translations do not use the word 'brain'. The actual word used in the original is *intelleto*. In Latin, *intellectus* meant perception or 'a perceiving'[17] and Symonds has, astutely in my view, rendered this as 'brain'.

Russian Suprematist, Malevich, who wrote that 'Art wants nothing further
to do with the object as such'[18]. I do not know what Malevich meant by
'as such' but I am pleased that he used the word 'further' in that statement, because vision, and in consequence art, cannot be divorced from
the objects in the world, a fact which also gives the lie to the Platonic
Ideal of an abstract form, existent in the world outside, independent of a
perceiving brain. For we know that people born blind because of a con-
genital cataract, and to whom vision is later restored, find very great difficulty in seeing. One of the first such operations was performed by a
French ophthalmologist on an eight year old boy[19]. The ophthalmologist
had anticipated the return of vision with much pride and enthusiasm.
'But', he wrote, 'the deception was great'. It took many months of training to teach his patient to recognize only a few objects by sight and these
he had forgotten within two years. Von Senden, who perfected the operation in Germany, tells us that many such patients are not even delighted
with the return of vision, if such it can be described, preferring to use the
sense of touch to recognize objects[20]. We now know that although the
connections from the eye to the brain are genetically determined, they
nevertheless atrophy and die if the visual system is not nourished during
a critical period after birth[21]. The consequence is that the cells in the
visual brain of an individual deprived of vision during the critical period
are either not visually responsive or respond in a vague and unpredict-
able manner, quite unlike the vigorous and selective responses of the
cells in the visual brain of a healthy individual. People deprived of their
vision have no Platonic Ideal of a couch or of anything else. Hence the
use of the word 'further' by Malevich at least implies that at one point
the artist did need the object, which is correct neurologically. So the
non-objective sensation and non-objective art of Malevich and his fol-
lowers is in fact the introspective art of a brain already well acquainted
with the visual world, with the objective world. It has a Platonic Ideal; it
has already selected all the essential information that is necessary for it
to identify and categorize objects. And true to its aims, of being a search
for essentials and constants, we find that as art developed more and
more in the modern era, it became better and better tailored to the physiology of the areas of the brain that I have mentioned, and specifically to
the physiology of the single cells in them, because the physiology of
these areas is itself tailored for extracting the essential information in the
visual environment. That is the beginning of the art of Malevich, which
is the precursor of much else in relatively modern art and which has,
without acknowledging its motives, tried best to explore this developed
inner neurological world, and has therefore tried to understand the
workings of the brain in a relatively simple and comprehensible way.

Not everyone, even among artists, admires modern art or some of the works I shall describe. John Ruskin, in his Friday evening Discourse given at the Royal Institution in February 1867, interestingly thought that the relatively modern art of the time was an expression of isolation of individuals in an alienated society. Other artists thought that such simplification debased art itself. Gustave Moreau, in whose studio Matisse worked, once told Matisse 'Vous voulez simplifier la peinture'[22], disapprovingly according to some and approvingly according to others. But Matisse, in a different context and time, replied:

> Underlying this succession of moments which constitutes the superficial existence of things and beings, and which is continually modifying and transforming them, one can search for a truer, more essential character, which the artist will seize *so that he may give to reality a more lasting interpretation* (my emphasis).[23]

No neurobiologist of vision could improve on this statement, save only to substitute the word 'brain' for the word 'artist'.

The art of the receptive field

What was the result of the so-called non-objective art of Malevich? A striking feature is the use of lines or bars of various shapes and widths and also of squares and rectangles (Fig. 7). In this he was followed by the Russian Constructivists, who also emphasized lines, as can be readily seen when one surveys the achievements of Russian artists of that period. They were not alone in this. Mondrian too ended by emphasizing the line but reached that end from a different beginning and with a different approach. He had started with naturalistic painting and had been much attracted to Cubism. But he had been disappointed with the development of Cubism, which had failed to get to the essence of form and had abandoned the quest, or so he believed. Instead it had developed in a different direction, characteristic of the later Synthetic Cubism, in which the emphasis was on the creation of new forms. Mondrian wrote that '... Cubism did not accept the logical consequences of its own discoveries; it was not developing abstraction towards its ultimate goal, the expression of pure reality ... To create pure reality plastically it is necessary to reduce natural forms to the *constant elements*' (original emphasis). 'Art', he thought, 'has two main human inclinations ... One aims at the *direct creation of universal beauty*, the other at the *aesthetic expression of oneself*'[24] (original emphasis). The first is more or less objective, the latter subjective. The first had to be objective because 'Since

Fig. 7 Suprematist painting by Kazimir Malevich. (By the permission of the Stede Lijk Museum, Amsterdam.)

art is in essence universal, its expression cannot rest on a subjective view', even if 'our human capacities do not allow of a perfectly objective view'. Art, he believed, 'shows us that there are also constant truths concerning forms' and it was the aim of objective art, as he saw it, to reduce all complex forms in this world to one or a few universal forms, the constant elements which would be the constituent of all forms, to '... discover *consciously or unconsciously* the fundamental laws hidden in reality' (my emphasis). This led him to pure abstract art, 'the art that is concerned with the basic elements of form'[25] and the search, through that art, led to the vertical and horizontal lines, or so he believed. These '... exist everywhere and dominate everything'. Moreover, the straight line, '... is a stronger and more profound expression than the curve'[24] because '... all curvature resolves into the straight, no place remains for the curved'[26]. He sought, in other words, the Platonic Ideal for form (though he did not describe it in these terms). He wrote, 'Among the different forms, we may consider those as being neutral which have neither the complexity nor the particularities possessed by natural forms or abstract forms in general'[24]. In this search, Mondrian settled on simple vertical and horizontal lines.

Malevich

Are these the intellectual wanderings of painters or were they really using non-objective sensation and experimenting with the capacities of the brain? I leave it to you to decide whether it was an accident, a mere coincidence, that the very elements, the lines, that Malevich thought represented non-objective art and that Mondrian thought represented the essentials of form and which both therefore emphasized so much in their work, are the very stimuli to which so many of the cells in specific areas of the visual brain are selectively responsive. These so-called orientation-selective cells, cells that respond to lines of certain orientation and not as well or not at all to lines of other orientations, are considered by physiologists to be the neural 'building blocks', or the essentials, of form perception (Fig. 8). Thus, in their separate searches, the different artists came to put on canvas the basic building block that the brain uses to represent forms. It is for this reason that I shall refer to this art, and to other kinds of art which can be related directly to the responses of single cells in the visual cortex, as '*the art of the receptive field*'.

The Métamalevich and the activation of V3

Cells that are specifically responsive to lines of particular orientation presented in their receptive fields are widely distributed in the visual brain. In some areas, like areas V3 and V3A, the cells respond best to lines of specific orientation if these lines are not held stationary within their receptive fields but actually move back and forth in a direction that is orthogonal to their preferred axis. It is perhaps this latter property that Gabo, Tinguely, and, more recently, Hugo DeMarco, have exploited in their work, though of course without reference to the brain. Jean Tinguely conceived of the neurologically interesting idea of setting the work of Malevich, dominated by oriented lines, into motion and calling the new creation the *Métamalevich* (or the *Métakandinsky* or *Métamatiques*). The prevalence of straight lines in motion would ensure that the *Métamalevich* is a strong stimulus for the cells of area V3 and V3A. It is interesting to note that the cells of both the latter areas are indifferent to colour, that is to say that they respond to lines of their preferred orientation irrespective of the colour of the oriented line; perhaps correspond-

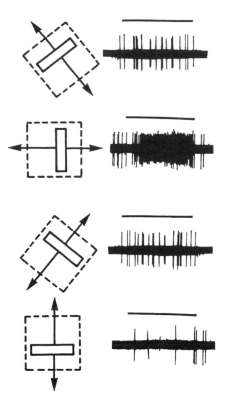

Fig. 8 This cell responds best to lines of a specific orientation, and responds less to other orientations.

ingly, Tinguely de-emphasized colour in his work, restricting himself to the use of blacks, whites, and greys; others, like Gabo in his *Kinetic Sculpture* and Hugo DeMarco in his *Série Relations* have done much the same. In fact, a very substantial restriction of the palette is very characteristic of kinetic art in general.

It is hard to think of visual stimuli that are better suited to activating the cells of the V3 complex (V3 and V3A) than the creations of Malevich and his successors and the extension of this work into the *Métamalevichs* of Tinguely. But the shift from a Malevich to a *Métamalevich* involves more than a shift in artistic form or emphasis. It actually involves the stimulation of a separate group of cells in the brain, and specifically in areas V3 and V3A. In both these visual areas, there are two groups of orientation-selective cells. One responds to oriented lines that are statically presented within their receptive fields, even in spite of the fact that the eye is continually moving back and forth in scanning them and thus generating a sort of passive motion of the stimulus. The cells of the other group can discount the displacement, and hence the movement, that is due to the eyes; its cells only respond if the oriented line itself is in actual motion, in other words it only responds to real motion[27].

Oriented lines also enclose squares and rectangles and both were considered to be basic forms by Mondrian—'the plurality of straight lines in rectangular opposition' as he called it. They were especially so considered by Malevich and Kandinsky, who thought that the square and the rectangle constituted the two most important elements of non-objective art. Many other artists, including Joseph Albers and the contemporary American artist Ellsworth Kelly as well as Ben Nicholson, emphasized these simple shapes in their paintings. Again, is it purely accidental that so many of the receptive fields of single cells are rectangular or square in shape? Take the receptive field of a single cell in area V4; this cell responds best to a blue patch against a white background (Plate 1) and does not respond much against a black background which in fact suppresses somewhat the cell's response. The resultant configuration of what this cell responds to best is not vastly different from a Malevich square (Plate 2). Is this a pure coincidence, or did Malevich and the others uncover a neurological secret? I leave this for you to decide.

Colour is a creation of the brain

There is a reason why the cell illustrated above, and others like it, have receptive fields with such sharp boundaries, and the reason is physiological, not artistic. This is perhaps best addressed by looking at the colour system, where the presence of a surround is critical in the determination of the colour of a given patch. Not surprisingly, it is a characteristic of area V4 that the cells in it have strong surrounds to their receptive fields, which influence their responses[28–30]. This is no doubt partly a reflection of the fact that the brain determines the colour of a surface by gauging the wavelength composition of the light reflected from it and comparing it with the wavelength composition of the light reflected from the surround. It is through this comparison that the brain is able to assign a colour to an area. This can be demonstrated by a simple experiment, which we owe to Edwin Land[31], but which is essentially a formal demonstration of what each one of us experiences many times during the day. It well illustrates the fundamental law of constancy, as applied to colour. If you were to arrange it so that one surface of a multicoloured abstract display containing no recognizable objects, say the green one, reflects a given amount of red, green, and blue light, say 30, 60, and 10 units respectively, and switch on all three projectors, you will perceive the colour of the green surface to be green. Hardly surprising, you may say, since in this condition the green patch reflects a lot more green light and the surround a lot less. But now if you arrange the same green surface to reflect 60, 30, and 10 units of red, green, and blue light respectively, that is twice the amount of red than of green light, and switch on all three projectors, the green area will still appear green. This is because, with the change from one illuminant to the other, the areas surrounding the green surface will reflect less green light and more red light and, since the brain is not interested in absolute energies but only relative values, it is able to take the ratio of the red and green light reflected from the centre and the surround and thus make itself independent of the absolute amounts. The same procedure applies to all the other squares of different colour.

Colour is therefore a comparison. That comparison is undertaken by the brain, not the world outside, and the result of that comparison belongs to the brain, not the world outside. It is one that is determined by the functional logic of the brain, nothing else. This leads me to agree with André Malraux when, in his book *Les Voix du Silence*, he qualifies as 'cette phrase maladroite' Cézanne's saying that 'Il y a une logique colorée; le peintre ne doit obéissance qu'à elle, jamais á la logique du cerveau' ('There is logic to colour; the painter must obey this, never the logic of the brain'). In fact, the construction of colour by the cerebral cortex[32] is an example of the brain going beyond the information given according to its own rules and logic. It was Newton who said that 'the Rays to speak properly have no Colour; in them there is nothing else than a certain power and disposition to stir up a sensation of this Colour or that'. The stirring up of that sensation requires the brain to undertake an operation—and thus go beyond the information provided by the physical environment.

The liberation of colour

Note that, to construct colour, the comparison that the brain undertakes is in the wavelength composition of the light reflected from one surface and the wavelength composition of the light reflected from surrounding surfaces. But the surrounding surfaces have a border with the surface in question, and the border has a shape. Hence the impossibility of separating colour from form. This, one is sometimes told, is what the fauvists tried to do. It is no wonder that none succeeded, because they tried something that is physiologically impossible. In the end, they opted for a different solution, which was to highlight the colour in a painting by attaching it to a form with which it is not normally associated, for example a red sea, or a green sun, and so on. Other artists who tried to highlight colour have also found it difficult to liberate it from form. The artistically uneducated brain could say with much validity that Rothko's creations consist very much of rectangles; the achromatopsic patient who cannot see colours will nevertheless still see the rectangles of Rothko, even though there is an attempt by Rothko to render form insignificant by repeating the same rectangle in different colours. The same is true of the creations of Robert and Sonia Delaunay; they may indeed be colourful but can equally be said to be composed of many rectangles and squares and simple shapes of various kinds. Hence, the simple shapes in the compositions of these painters can be differentiated because of colour or because of shape and in practice because of both. Perhaps the easiest way of emphasizing colour is to remove the simple shapes which are so essential a part of the brain's physiology, as Malevich instinctively understood, and create a garble of nonsense in colour, depriving the painting of even a title, and hence of a cognitive element. The man, I believe, who came closer to achieving this is Wilhelm de Kooning, and it is noteworthy that a significant number of his later, form-wise nonsensical paintings are untitled. And just as Sartre praised the mobiles of Calder for signifying nothing, so the lack of a discernible shape in de Kooning's paintings means that they signify nothing and emphasize colour because of that.

Kinetic art and the physiology of area V5

It is interesting to consider further the relationship of single-cell physiology to visual art and the assumption that I have made that artists are, unknowingly, exploring the organization of the visual brain, though with techniques unique to them. Kinetic art provides fertile ground for doing so[33]. This is an art in which actual motion is an integral part of the work. It started as a dissatisfaction with an art that seemed to exclude movement, or what Naum Gabo called the 'fourth dimension'. The first steps taken to remedy this were hesitant and consisted of representing motion statically. They are well exemplified in the works of Marcel Duchamp (whom Etienne described as the 'Frenchman who engages himself in dissecting sensations and sentiments'). There is little doubt that Duchamp's creations were strongly influenced by the chronophotography of Jules-Etienne Marey in France (Fig. 9) and Edward Muybridge in England. A succession of paintings, like *Dulcinea* and *Nu descendant l'Escalier II* (Fig. 10) are strongly suggestive of movement and show

Marey's influence. Of the latter, Duchamp wrote that it 'was the con-
vergence in my mind of various interests amongst which the cinema,
still in its infancy, and the separation of static positions in the photo-
chronographs of Marey ... the anatomical nude does not exist, or at least
cannot be seen since I discarded completely the naturalistic appearance,
keeping only the abstract lines of some twenty different static positions
in the successive act of descending'[34]. A similar influence had taken
hold among the Futurists in Italy, who had declared the importance of
speed in their *Manifesto of Futurism*. But, here again, for all its declared
importance, motion was represented statically in the paintings of Boccioni,
Balla, and Russolo. Some, like Ettore Bugatti, disappointed by the lack of
motion in works of art, abandoned painting altogether and took refuge in
designing fast cars, themselves often works of considerable beauty.

This static representation of motion was soon changed by the work, not
of those who proclaimed its importance, but of one who was fascinated
by seeing motion. It was the Swiss artist Jean Tinguely who became
much intrigued with motion at an early stage in his career, ever since he
saw Georges Mathieu paint. He tells us that his fascination was not with
the finished product but with the movement while Mathieu was executing the painting. He thus decided to make movement an integral part of
the work of art. In his *Métamalevichs* and *Métamatiques* (Fig. 11) he had
simplified forms and greatly restricted his palette, even eliminating
colour altogether, while highlighting motion; he thus produced paintings
which are wonderful stimuli for the cells of areas V3 and V3A. But there
is another area in the visual brain, area V5, whose cells are even more
selective for motion. Most of these are directionally selective, that is to
say responsive to motion in one direction but not in the opposite direc-

Fig. 9 *Long-jumper* by Julés Ettiene Marey. (By permission of
the Musée Marey, Beaune.)

Fig. 10 *Nu descendant l'Escalier II* by Marcel Duchamp. (Philadelphia Museum of Art; The Louise and Walter Arensberg Collection.)

tion (Fig. 2), and they respond best to spots rather than to lines moving in the appropriate direction (Fig. 12). Like the cells of V3, those of area V5 are indifferent to the colour of the stimulus, responding to their preferred direction irrespective of the colour; they are mostly indifferent to the form of the stimulus too. V5 is, therefore, an area in which motion is

Fig. 11 *Metamecanique* by Jean Tinguely.

emphasized and both colour and form are de-emphasized or rendered irrelevant[35].

Given its specialization, it is not surprising to find that when V5 is destroyed by a lesion, the patient is not able to see objects when in

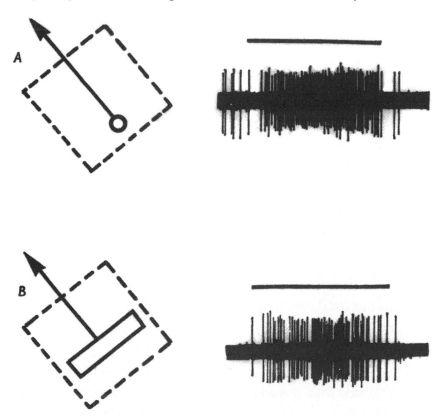

Fig. 12 This cell responds more vigorously to spots than to bars passing through its receptive field. (From Zeki, S., *J. Physiol.*, 1974; **236**, 549–573.)

motion but only when they are stationary[36-38]. Such a patient has difficulty in seeing cars in motion, or to see tea being poured because of the inability of seeing the level rise, or indeed to carry on a conversation with the same ease as normals because of their incapacity to see lips move. She is, of course, also unable to appreciate kinetic art. This is not to imply that the aesthetics of kinetic art is mediated through area V5, but only that V5 is necessary to it.

To best tailor a stimulus to the physiology of area V5, the emphasis had to be on motion, with both form and colour de-emphasized, or rendered meaningless. It was Alexander Calder who came closest to achieving this, by inventing his mobiles (Fig. 13), which represent the high point of kinetic art. Calder apparently hit upon the idea after visiting Mondrian's studio. This is surprising because the nearest Mondrian ever got to putting motion in his paintings was in his static *Broadway Boogie-Woogies*, where the motion is only suggested by the name, not much else. However that may be, Calder genuinely hit on the idea of the mobile, small roundish objects that move in different directions. As seen from a distance, these are remarkably effective stimuli for the cells of area V5. Remember that the cells of area V5 are indifferent to colour, that is to say that they will respond to a stimulus in motion regardless of its colour. In the laboratory, we usually stimulate the cells of area V5 with white spots against a black background or the reverse. And Calder, just

Fig. 13 *White Mobile with 24 Pieces* by Alexander Calder. (Private collection, France.)

like Tinguely before him, began to make his mobiles achromatic, that is to say in black and white. He considered that the other colours, with the possible exception of red, 'confused' the clarity of the mobiles. What does 'confusing' the clarity of mobiles mean in neurological terms? We have found that, in general, when we stimulate with colour, activity in area V4, specialized for colour, goes up and activity in area V5, specialized for motion, goes down (Fig. 14). So maybe, without realizing it, Calder was uttering a neurological fact about the brain, which we have just begun to discover with neurological tools.

Area V5 is situated laterally and ventrally in the human brain. It provides fertile ground for testing the proposition that the brain actively generates percepts and is not a mere passive chronicler of outside events. It is interesting to see what happens when we perceive something in a work of art which is not objectively there using V5. An example is to be found in the work of the contemporary French physiological artist, Isia Leviant. In his *Enigma*, some (though not all) of you will perceive rapid motion confined to the rings. If we were to ask subjects who can see the motion in the rings to look at *Enigma* and then measure the activity in their brains (Fig. 15), we shall find that it is largely confined to area V5. When the same subjects look at objective motion, we find that there is activity in both V5 and V1. Hence it is as if activity in V5 is imposing certain properties on *Enigma*, properties which are not objectively there. The brain thus goes beyond the information given, and constructs the image according to its own rules.

I have tried to show you that we have gained sufficient knowledge about the visual brain in the past twenty-five years to be able to say something both useful and interesting about what happens in the brain when we look at works of art and to consider visual art as an extension of the functions of the visual brain. That function is the acquisition of knowledge and the self-confessed aim of art, from artists and philosophers, is also the acquisition of knowledge. We may therefore regard the aim of art as nothing more nor less than an extension of the aims of the visual brain. There are of course many areas that we have not even started to explore yet: the powers of art to disturb and arouse, the role of the imagination in generating new works of art and the relationship of all art, since it is a self-reproductive process, to the atavistic impulses of sexuality. As well, we have no idea of why some works of art are more appealing aesthetically to some people than to others or why different painters are drawn to executing different kinds of work.

To many the notion of talking about art in the relatively elementary and yet precise terms of what happens physiologically in the brain may seem a somewhat dangerous thing to do. It implies that what happens in

Increases in rCBF

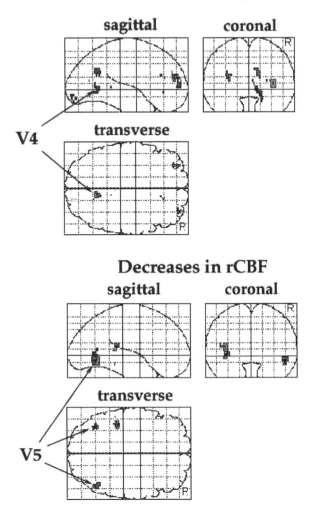

Fig. 14 Activation in area V4 leads to a decrease in activity in area V5.

the brain of one perceiver is very much the same as what happens in the brain of another. Art, they might argue, is an aesthetic experience whose basis remains opaque and mysterious, unqualified by scientific experimentation and dissection, and indeed should continue to remain so; physiology, they might argue, should above all not assault the closeted secrets of fantasy. To reduce art thus to a physiological formula etiolates art itself which, many would argue, has gained a great deal of its value and appeal by its ambiguity, the different way in which it nourishes,

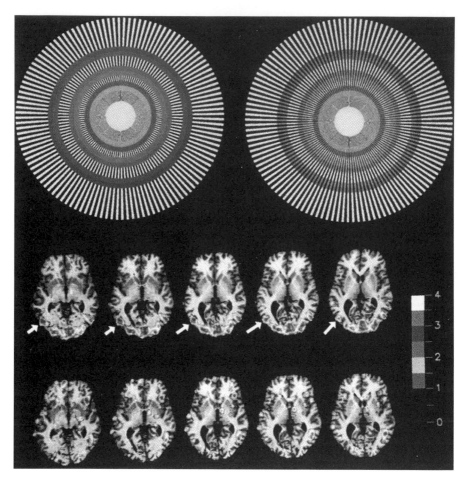

Fig. 15 When most people view the figure on the left, they see an enigmatic, circular motion in the grey rings. This illusion can be destroyed by making the spokes intersect the rings. This can provide a stimulus for a PET scanner. When subjects placed in the scanner are shown the two images, and the activity pattern of the control is subtracted, increased activity is detected in an area very close to the area V5 detected using moving dots. (Results from Zeki, S. *et al.*, *Proc. R. Soc. Lond. Biol.*, 1993; **252**, 215–222.

arouses and disturbs different individuals. The supposition of a basic similarity between different individuals in so subjective an area, and the profaning of the secrets of the brain in this way may, they fear, lead to a world where none would be allowed the privileges of concealment and privacy and subtlety, where we would all be mentally naked together since we would all be aware of what happens in one anothers' brains.

I would argue precisely the opposite. It is true that, to a large extent and perhaps at an elementary level, what happens in one brain is very similar to what happens in another, which is one reason why we can talk about art and communicate through it and why such works as Michelangelo's *Piéta* or Botticelli's *Primavera* have had such a universal appeal. Nor does knowledge about what happens in the visual brain threaten the aesthetic sense. Just as knowledge of the fact that the motion in the rings of the *Enigma* illusion is a product of activity in a highly specific area of the visual brain does not in any way interfere with the perception of that motion, so no physiological or anatomical dissection of the brain, no profound knowledge of it, can take away our perception of the resentment at failing powers in Rembrandt's later self-portraits, or of our perception of the world of loneliness and misery and pain that Edvard Munch captured in *The Scream*, or the world of isolation and depravity that Degas portrayed in *Absinthe*, or of our perception of the serenity in the face of Michelangelo's Christ, after the supreme doubt expressed in the last seven words from the Cross. The evocative power of these paintings is prodigious, being itself the product of the prodigious powers of the brain. Indeed, because of their almost infinite variety, the imagination of the brain, the painter's as well as ours, is impossible to capture in a single painting or indeed many single paintings. I think that the mighty Michelangelo, who tried so hard to portray the spiritual beauty of Christ, especially in his last moments, realized this instinctively towards the end of his life and, embittered, understood the limits of what painting and visual art could achieve in comparison with the seemingly effortless and instantaneous power and success of the brain. It is perhaps for this reason, among others that art historians ascribe to it, that in one of his last and most pathetically beautiful sonnets, dedicated to Giorgio Vasari, he wrote:

> I now know how fraught with error was
> the fond imagination which made
> Art my idol and my king
>
> No brush, no chisel, will quieten the soul
> Once it is turned to the divine love of Him who, upon the Cross,
> outstretched His arms to take us unto Himself

And no knowledge of the brain and its intricate operations can quieten the soul when it begins to contemplate how the many billions of neurons that constitute it work in unison to create works of such infinite beauty but whose beauty pales into insignificance compared with the immense beauty of the brain itself.

References

1. S.M. Zeki, *Nature*, 1978, **274**, 423.
2. S. Zeki, J.D.G. Watson, C.J. Lueck, K.J. Friston, C. Kennard, *et al. J. Neurosci.*, 1991, **11**, 641.
3. S. Zeki, *Brain*, 1990, **113**, 1721.
4. C.A. Pallis, *J. Neurol., Neurosurg. Psychiatry (London)*, 1955, **18**, 218.
5. F. Lhermitte, F. Chain, R. Escourolle, B. Ducarne, and B. Pillon, *Rev. Neurol. Paris*, 1972, **126**, 329.
6. L. Campbell, *Renaissance Portraits: European Portrait-painting in the 14th, 15th and 16th Centuries*, Yale University Press, New Haven, Connecticut, 1990.
7. D. Tranel, A.R. Damasio, and H. Damasio, *Neurology*, 1988, **38**, 690.
8. S. Zeki, *A Vision of the Brain*, Blackwell, Oxford, 1993.
9. H. von Helmholtz. *Handbuch der Physiologischen Optik*, Vol. 2, Voss, Hamburg, 1911.
10. A. Gleizes and J. Metzinger. *On Cubism*. Fisher Unwin, London, 1913.
11. Plato, *The Collected Dialogues*, Princeton University Press, Princeton, New Jersey 1961.
12. H. Read, *The Philosophy of Art*, Faber and Faber, London, 1964.
13. A. Schophenhauer, The World as Will and Idea (1844) In *Philosophies of Art and Beauty* 3rd book (ed. A Hofstader and R. Kuhns), University of Chicago Press, Chicago 1976.
14. J. Constable, *Syllabus of a Course of Lectures on The History of Landscape Painting*, Royal Institution of Great Britain, London, 1836.
15. N.K. Logothetis, J. Pauls, and T. Poggio, *Curr. Biol.*, 1995, **5**, 552.
16. J. Rivière (1912), *Present Tendencies in Painting*, reprinted in *Art in Theory 1900–1990. An Anthology of Changing Ideas* (ed. C. Harrison and P. Wood), p. 183, Blackwell, Oxford.
17. R.J. Clements, *Michelangelo's Theory of Art*. New York University Press, New York, 1961.
18. K. Malevich (1919), *Non-objective Art and Suprematism*, reprinted in *Art in Theory 1900–1990. An Anthology of Changing Ideas* (ed. C. Harrison and P. Wood), pp. 290–92, Blackwell, Oxford.
19. Moreau *Ann. Oculist.* (Paris), 1913, **149**, 89.
20. M. von Senden, *Raum- und Gestaltauffassung bei Operierten Blindgebornen*. Methuen & Co., London, 1932.
21. D.H. Hubel and T.N. Wiesel, *Proceedings of the Royal Society (London)*, 1977, **B198**, 1.
22. S. Zeki and Count Klossiwski de Rola Balthus, *La Quéte de l'Essentiel*. Les Belles Lettres, Paris, 1995.
23. H. Matisse, (1908), *Notes of a Painter*, reprinted in *Art in Theory 1900–1990. An Anthology of Changing Ideas* (ed. C. Harrison and P. Wood), p. 72, Blackwell, Oxford.
24. P. Mondrian (1937), *Plastic Art and Pure Plastic Art*, reprinted in *The New Art—The New Life, The Collected Writings of Piet Mondrian* (ed. H. Holtzman and M. James), pp. 288–300, G.K. Hall & Co., Boston.

25. P. Mondrian (1929), *Pure Abstract Art*, reprinted in *The New Art—The New Life, The Collected Writings of Piet Mondrian* (ed. H. Holtzman and M. James), pp. 223–225, G.K. Hall & Co., Boston 1987.

26. P. Mondrian (1919), *Dialogue on the New Plastic*, reprinted in *The New Art —The New Life, The Collected Writings of Piet Mondrian* (ed. H. Holtzman and M. James), pp. 75–81, G.K. Hall & Co., Boston 1987.

27. C. Galletti, P.P. Battaglini, and P. Fattori, *Exp. Brain Res.*, 1990, **82**, 67.

28. S. Zeki, *Neuroscience*, 1983, **9**, 741.

29. R. Desimone and S.J. Schein, *J. Neurophysiol.*, 1987, **57**, 835.

30. S. Zeki, *Nature*, 1980, **284**, 412.

31. E. Land, *Proceedings of the Royal Institution of Great Britain*, 1974, **47**, 23.

32. S. Zeki, *Proceedings of the Royal Institution of Great Britain*, 1984, **56**, 231.

33. S. Zeki and M. Lamb, *Brain*, 1994, **117**, 607.

34. M. Duchamp (1912), reprinted in *Duchamp Retrospective at the Palazzo Grassi, Venice 1993* (ed. G.E. Fabbri), Gruppo Editoriali Fabbri, Milan.

35. S.M. Zeki, *J. Physiol.*, 1974, **236**, 549.

36. J. Zihl, D. Von Cramon, and N. Mai, *Brain*, 1983, **106**, 313.

37. S. Shipp, B.M. de Jong, J. Zihl, R.S.J. Frackowiak, and S. Zeki, *Brain*, 1994, **117**, 1023.

38. J. Zihl and J.C. Mayer, *Nervenarzt*, 1981, **52**, 574.

SEMIR ZEKI

Born 1940, Professor of Neurobiology at University College London, where he started his studies in medicine and then abandoned this in favour of a scientific career, terminating his formal training with a PhD in Anatomy. He did his postdoctoral work in the United States (Washington DC and Wisconsin) before returning to England to resume his work on the functional organization of the primate visual brain. He was Henry Head Research Fellow of the Royal Society from 1975 to 1980 and became Professor of Neurobiology in 1981. He is a member of the Academia Europaea and the European Academy of Sciences and Arts. His main interest lies in understanding how the visual brain works, an area in which he has published many specialized articles as well as a more general book entitled *A Vision of the Brain*. He has also maintained an interest in visual art, especially in relation to the visual brain, publishing articles on the subject. His most recent contribution in this area is a series of conversations on the subject with the French painter Balthus, published this year under the title of *La Quête de l'Essentiel*.

Exploring the Universe with the Hubble Space Telescope

ALEC BOKSENBERG

The astronomical context

Imagine a clear, moonless night in the country. We can see many thousands of stars. Light from some of these set out hundreds or thousands of years ago and they appear to us as they were then. In a great arc across the sky shines the Milky Way which Galileo showed to be composed of myriads of faint stars. This is our Galaxy, a discus-shaped city of more than 100 000 million stars interspersed with gas and dust, all held together by gravity and rotating in a spiral form.

It was Edwin Hubble in the early 1920s who showed that the apparently small spiral nebulae others had thought were inside our Galaxy were themselves similar star systems at great distances. The Andromeda Galaxy, two million light years away, is the nearest of these. (A light year is the distance light travels in a year: nearly 10 million million kilometres.) It closely resembles our Galaxy and like ours it has dwarf companion galaxies hovering nearby. It appears to the naked eye as a faint smudge but its structure can be seen with a good pair of binoculars. Galaxies in general come in a variety of shapes and are often found associated in groups or in larger clusters. The Milky Way Galaxy is itself part of a loose collection of galaxies located far outside a rich cluster, the Virgo Cluster, containing several thousand galaxies.

Again, it was Edwin Hubble who found that the clusters of galaxies were receding from us, in such a way that everything in the Universe was moving apart from everything else in proportion to distance: the further away a galaxy cluster, the faster it is receding. If we think of raisins in a cake baking in the oven we recognize, raisin to raisin, the same phenomenon occurring. There is no unique point in this expansion; rather, the whole cake—that is the whole of space—is expanding.

We can trace back the expansion of the Universe to a time about 15 billion (15 thousand million) years ago, when all the galaxies we can see within our horizon were on top of one another. It was then that the cosmic expansion was initiated.

As we have seen, this expansion was not like an explosion familiar on Earth, starting from a definite centre and spreading out into the surrounding space, but one which occurred simultaneously everywhere, filling all space from the start, with every particle of matter rushing apart from every other. In its early phase the Universe was unimaginably hot and dense. As it expanded it cooled. A hundredth of a second after what has become known as the Big Bang it was at a temperature of about 100 000 million degrees, which is much hotter than at the centre of even the hottest star today; so hot, in fact, that none of the components of ordinary matter—molecules or atoms or even the nuclei of atoms—could have held together. Only minutes later some nuclear reactions occurred. Essentially all of the helium nuclei in the Universe we see today, with traces of deuterium and lithium, were synthesized at that time; and hydrogen then, as now, was the dominant form of matter. As the Universe cooled further, this primordial gas eventually condensed into galaxies of stars arranged in vast aggregations. In the stars were synthesized heavy elements such as carbon, oxygen, silicon, and iron, and some of this material was ejected more or less violently into the interstellar medium. In turn, more stars formed from clouds of this enriched material. It was only then that chemically complex structures like ourselves became possible.

The telescope in orbit

Much of what we know about the Universe has come from observations made over many decades with powerful telescopes equipped with sophisticated instruments, operating at excellent mountain-top sites remote from city lights and pollution. It says a lot, therefore, for the capability and performance of the Hubble Space Telescope (Fig. 1) that in its relatively short life it has made unique and very substantial contributions to astronomical knowledge. It is not surprising that this is so, for the construction and operation of the Hubble Space Telescope, jointly between NASA and the European Space Agency, is one of the most ambitious (and most expensive) astronomical projects ever conceived.

By ground-based standards the Hubble Space Telescope is not a large telescope. The diameter of its primary mirror is 2.4 metres, giving it just one-sixteenth the collecting area of the largest available ground-based

Fig. 1 Artist's impression of the Hubble Space Telescope in orbit. The two panels carry solar cells which provide electrical power. The open flap shields the mouth of the telescope from sunlight. (Reproduced with permission from NASA.)

telescope. Yet from its vantage in orbit it has a performance no mountain-top telescope can match, whatever its size. Observing from space brings three key benefits. First, there is no blurring from atmospheric turbulence. Figuratively put, star-watching through kilometres of air is like bird-watching from the bottom of a swimming pool. Second, without atmospheric absorption the deep ultraviolet region of the spectrum becomes available. Much of our information about hot stars and the inter-stellar medium comes from this region. Finally, the lack of atmospheric emitted and scattered light gives a darker sky, which greatly increases the relative contrast of the faintest objects.

To gain the most from these advantages, the all-important primary and secondary mirrors and the telescope pointing control were made to specifications of unprecedented accuracy. The goal was to concentrate most of the visible light from a star into a circle only 0.05 arc seconds across. This is equivalent to the diameter of a pound coin as it appears from a distance of 100 kilometres. In contrast, the best star images recorded from the ground rarely come as small as ten times this size. Added to this, a host of other aspects of design and almost countless items of detail were incorporated to realize a practical, efficient, and reliable observatory. This would need to satisfy the astronomers from the partner nations (any of whom could compete for observing time by submitting research proposals) over a specified operating life of 15 years. As a measure to maintain this long operating life, frequent servicing visits by space shuttle crews to carry out maintenance and repairs in orbit were specified from the outset. As it turned out this perceptive piece of planning was crucial in achieving a successful outcome for the project.

The telescope, however, is nothing without instruments to receive and analyse the images it delivers. Just as in ground-based telescopes, the instruments on the Hubble Space Telescope are mounted behind the primary mirror. There, protected by the spacecraft's cavernous aft shroud, lie two cameras—the Faint Object Camera and the Wide Field and Planetary Camera—and two spectrographs—the Faint Object Spectrograph and the High Resolution Spectrograph. Light from the sky enters the forward telescope tube, is reflected back from the concave primary mirror at the tube's base, leaving it in a converging beam until it meets the small convex secondary mirror set axially near the mouth of the tube, then is reflected back to the centre of the primary mirror still converging but less steeply so. It passes through a hole in the primary mirror to be focused onto one or other of a set of aperture plates each of which relates to one of the instruments. Any of the instruments can be brought into use by pointing the telescope so that the image of a target field enters the appropriate aperture.

The Faint Object Camera was built by the European Space Agency to a design which centred on the use of an extremely sensitive camera device known as the Image Photon Counting System which I had developed for ground-based telescopes. In this instrument, individual photons of light are detected and recorded as an accumulating image in a computer memory. Of all the instruments, the Faint Object Camera can detect the faintest objects in the ultraviolet and much of the visible region and achieve the highest angular resolution. The Wide Field and Planetary Camera uses charge-coupled device detectors, whose sensitivity extends to the near infrared region, in an optical system that gives a wider field but at lower resolution than does the Faint Object Camera. The spectrographs—these spread out the light of individual objects or small regions of extended objects into a 'rainbow' of colour—have complementary performance, with the Faint Object Spectrograph able to examine fainter objects but at lower spectral resolution than the High Resolution Spectrograph. Spectroscopy is the most commonly applied analytical technique in astronomy. It is used to determine chemical composition and abundance by recording the spectral signatures of atomic and molecular species, to deduce physical properties including temperature and gas pressure and, through the Doppler effect, to measure bulk radial velocity or breadth of velocity distribution due to gas temperature or turbulence.

The Hubble Space Telescope was released into its orbit from the bay of the space shuttle *Discovery* in April 1990. The initial euphoria quickly evaporated with the discovery that the main mirror was the wrong shape: it was too flat by an amount equal to about one fiftieth the diameter of a human hair. Although this might seem small, it is a gross mistake by the standards of modern precision optics. Instead of the image of a star being tightly focused, most of the light was spread into a fuzzy halo. After a period of—to put it mildly—disappointment, it was taken as a monumental challenge to find a way out of this apparent disaster. Astronomers quickly learned how to wring as much information as possible out of the aberrated images. Computer processing was used to enhance the images and much of the lost resolution quality was recovered but at a cost of using only about 20 per cent of the light, so the telescope was very inefficient. Even with this disadvantage much useful astronomy was done, but to realize its full potential the telescope would have to be repaired. The large mirror itself could not be altered or exchanged. What was devised, therefore, was a means of correcting for the telescope's impaired vision much in the same way as eyeglasses correct human sight. A new Wide Field and Planetary Camera, with correcting optics built into it, and a set of small correcting mirrors using mechanical arms to position them in front of the other three instruments, were produced in remarkably short time and installed in December 1993 by the crew of space shuttle *Endeavour* (Fig. 2). To make the most of the occasion many maintenance and other repair tasks were included in the crew's programme of work. When they returned to Earth 11 days later everything planned had been accomplished and, as was soon proved when the first images appeared, the mission had been a complete success. In fact the telescope's superb imaging performance (Fig. 3) now exceeded the original very stringent specification. The second servicing mission is planned for 1997 when two new instruments will be installed (and two removed) and another set of maintenance and repair tasks will be done.

The Hubble Space Telescope is used around the clock without a break every day of the year. After the competitive selection of observing proposals, the programmes awarded time are scheduled into an optimized sequence to be executed automatically. Step-by-step instructions for the telescope then are produced from this sequence and are sent up from the control centre near Washington via a relay satellite. The completed observations are relayed back to the control centre for processing and delivery to the observers.

In the remainder of this article I will give some examples of the remarkable results obtained with the Hubble Space Telescope. Necessarily these will be few in number, but the scientific output of the telescope is vast. To do full justice to the images I show, I highlight them against a descriptive backdrop of the scientific fields they help to advance.

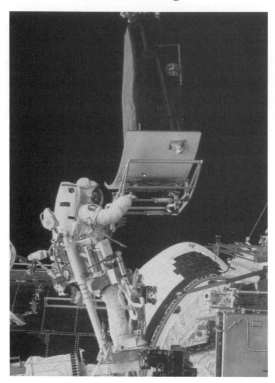

Fig. 2 A crew member of the space shuttle *Endeavour* handling instrumentation from the Hubble Space Telescope during the installation of correcting optics in December 1993. (Reproduced with permission from NASA.)

The Solar System

Although it was designed for the observation of some of the most distant objects in the Universe, the Hubble Space Telescope has proved well suited to studies of bodies in the Solar System. Detailed images of the planets, for example Jupiter, Saturn, and Mars, have been valuable in showing long-term changes in weather patterns and surface features not possible from the ground because of the lack of clarity, or from the *Voyager* space probes which could spend only a short time near each planet visited.

A particular spectacular event studied with the Hubble Space Telescope was the impact of Comet Shoemaker–Levy 9 with Jupiter in July 1994. The comet was a strange object made up of about 20 individual fragments moving in formation. The fragments had come from a large comet that was torn apart by Jupiter's gravitational pull during a very close approach of the comet in 1992, and were now returning. Although

Fig. 3 Comparison images of stars in the same small region of sky obtained with a ground-based telescope in best atmospheric conditions (left) and the Hubble Space Telescope before (middle) and after (right) correction. (Reproduced with permission from NASA.)

the successive impacts occurred on the side of Jupiter facing away from the Earth, in some of these material was sent high enough into Jupiter's atmosphere to be seen only a few minutes after impact. When the impact zones rotated into view there were dark scars where the fragments had entered and disintegrated in huge explosions. The early development of some of the impact zones is displayed in Plate 4. First, there is hint of one of the impact plumes, then a view of the corresponding fresh impact site, then the evolution of this and other nearby sites in the wake of atmospheric winds. For months afterwards the evolution of the impact scars was followed as Jupiter's winds stirred up the debris raised by the impacts, giving much information about Jupiter's dynamic atmosphere.

Jupiter has several moons, four of them discovered by Galileo. One of these, Io, is the Solar System's most dynamic moon. It orbits rapidly very close to Jupiter, once every 1.8 Earth days, and is forced into a dynamic resonance between Jupiter and two other of the Galilean satellites, Ganymede and Callisto. The strong gravitational influences on Io lead to gross flexing which causes heating and melting. Eventually, the molten interior reaches the surface and is expelled in volcanic sulfur-rich flows and plumes. In Fig. 4 a pair of Hubble Space Telescope images of Io shows evidence of violent volcanic activity from the emergence of a 300 kilometre diameter yellowish-white feature which arose during the period of 16 months between the two observations. The realization that from the ground Io appears little more than a point of light drives home the astonishing level of detail seen in these views.

Stars and the interstellar medium

The Universe at large seems to be mostly empty. There are vast reaches of open space between the stars in our Galaxy and it is clear that the

Alec Boksenberg

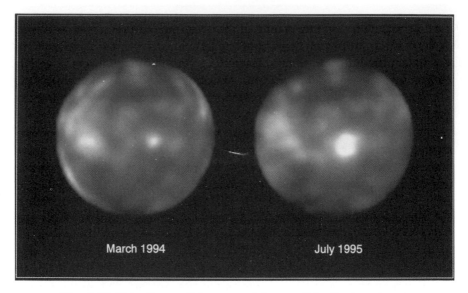

March 1994

July 1995

Fig. 4 Hubble Space Telescope pictures of Jupiter's volcanic moon, Io, showing changes which occurred during a period of 16 months. (Reproduced with permission of STScI and NASA.)

same is true in other galaxies. There are enormous distances between the galaxies. But it is a fact that nowhere is the Universe quite empty. At least a slight trace of matter is present everywhere in space, although mostly it is far more sparsely filled than the best 'vacuum' which can be produced in the laboratory.

The main constituents of interstellar matter are everywhere the same—about 75 per cent hydrogen and 25 per cent helium by mass— but the proportions of the minor constituents differ widely between locations. The total density and other properties of this matter also differ widely from place to place. Mostly the matter is gaseous but a small proportion is in the form of minute, solid dust particles. On average there is only about one atom per cubic centimetre and one dust grain per hundred thousand cubic metres of space. Although the density of interstellar matter is extremely low, the volume of the space in a galaxy is so great that the total quantity of interstellar material is very considerable. Our own Galaxy contains a mass equal to about 10 billion times the mass of the Sun in gaseous material between the stars, making up about 10 per cent of its total apparent mass. Most of this matter is distributed in the spiral arms and the disk of our Galaxy and is confined to a layer about a thousand light years thick.

There is a continuing exchange of matter between the stars and the interstellar medium within the disk. First, stars condense out of the primordial material which ultimately forms into galaxies, although it is not yet clear when in the early life of a galaxy this occurs. The less massive stars live on for many thousand millions of years and die relatively quietly. The very massive stars, many times more massive than the Sun, burn out after only a few million years and end their lives in a brilliant supernova explosion.

A star is a ball of gas in balance between the inward pull of its own gravitation and the outward pressure of the hot gas within. Stars form from clouds of gas and dust (if yet present) in the denser regions of the interstellar medium. When the density of a cool cloud exceeds a critical value it can collapse upon itself gravitationally, building up the temperature and pressure enormously in the interior. Initially the collapse is comparatively rapid, taking about a million years, and a protostar is formed.

For a star like the Sun the contraction then proceeds more slowly until the centre becomes hot enough to ignite nuclear reactions. Eventually it reaches a steady condition, converting hydrogen nuclei to helium nuclei in its core and adding to the helium which has been made in the Big Bang. The energy yield is extremely large and there is an ample supply of fuel in the core for a total lifetime of 10 billion years; the Sun is about halfway through this steady period. The temperature at the centre is then 15 billion degrees. Heat is transported out by radiation or convection, in different radial zones, and is radiated from the surface which in the Sun maintains itself at an equilibriunm temperature of about 6000 degrees. When, eventually, most of the central hydrogen is used up, the core of helium 'ash' contracts under its own gravity and grows hotter. Fusion moves outwards to a shell surrounding the core where hydrogen-rich material is still present. During this phase the outer layers of the star, responding to the greater internal pressure, swell enormously, transforming it to an outwardly cool but very luminous 'red giant' star. When this happens to the Sun its radius will extend beyond the Earth to the orbit of Mars. A star of the Sun's mass endures as a red giant for only a few hundred million years. The helium core continues to heat until new nuclear reactions arise with the helium itself becoming a fuel, forming carbon and oxygen nuclei. In the last stages of the star's life it becomes unstable and its outer layers are pushed off in a shell of glowing gas, still illuminated by the bright core it has left behind. William Herschel, noticing such faintly gleaming clouds, called them planetary nebulae, wrongly guessing that these are young objects that had not yet condensed into stars. Nuclear reactions now cease and the bright remnant core, known as a 'white dwarf', cools off slowly, over billions of years, prevented from collapsing further by the pressure from the quantum repulsion of its free electrons, while keeping its fusion products locked up inside it—a dying stellar ember.

An impressive image of the planetary nebula NGC 6543, nicknamed the 'Cat's Eye' Nebula, obtained with the Hubble Space Telescope (Plate 5), reveals surprisingly intricate detail and structures not visible from the ground. Lying 3000 light years away, it is one of the most complex planetary nebulae observed. The light we now see left it when it was at least 1000 years old. The nebula glows in the light initiated by the ionization of the gas by ultraviolet photons from the very hot white dwarf star. Visible light is produced dominantly when electrons subsequently are recaptured by hydrogen or helium ions or following the collision of electrons with ionized oxygen, neon, or nitrogen atoms.

Were this sequence of events true of all stars, the Universe would be a duller place—and we would not be here to comment on it. However, stars more massive than the Sun lead a shorter but more intense life. A star of 20 solar masses starts with a higher central temperature and pressure than the Sun, is 20 000 times more luminous, rushes through its hydrogen-fusing phase a thousand times faster and swells to a red giant after just 10 million years. The high core temperature also leads to a more diverse set of nuclear reactions, with heavier elements successively produced in a developing central 'onion-skin' structure, where the inner, hotter shells 'cook' elements further up the periodic table, all the way up to iron. Then a catastrophe occurs. In fusion reactions nuclei lighter than iron release energy, but to make a nucleus heavier than iron requires an intake of energy, so once a star has built an iron core it can no longer generate energy by fusion. Suddenly—in the space of just one second— the core collapses and becomes as dense as a nucleus. More material falls onto this now incompressible core, leading to an intense shock wave which propagates outwards to the star's surface; the star suddenly brightens and explodes as a supernova. For some weeks the star shines as brightly as a billion Suns, while the surface expands at several thousand kilometres a second. Not only are the accumulated interior fusion products ejected violently into the neighbouring gas, but new elements are synthesized in the intense heat behind the outgoing shock wave, now transcending the iron 'barrier' and producing elements all the way up to uranium. Among the many complex reactions occurring, free neutrons bombard iron nuclei and produce gold; gold is then transformed into lead—an alchemist's nightmare!

In February 1987 a supernova, SN 1987A, occurred 165 000 light years away in one of our companion galaxies, the Large Magellanic Cloud, and subsequently was studied extensively with the Hubble Space Telescope. Images obtained in 1994 revealed an astonishing system of rings around the supernova (Plate 5). The inner ring is material which the star lost when it was a red giant, subsequently excited by the flash of ultraviolet light from the supernova. The outer rings are more mysterious but also must relate to gas lost by the star before the supernova stage. The exploding shell of the star carrying the nuclear products is expanding rapidly but so soon after the event still is relatively compact. It is seen near the centre of the rings structure in Fig. 5. The expanding debris of past supernova events can be seen permeating large regions of our Galaxy. A small portion of one such system, called the Cygnus Loop, 2500 light years away, has been imaged with great clarity with the Hubble Space Telescope (Plate 6); this beautifully intricate glowing structure of rapidly expanding gas loaded with heavy elements, and impacting the interstellar gas, has been flung to a great distance from a supernova which occurred 15 000 years ago. In this way massive stars return to the interstellar medium much of the hydrogen and helium from which they were formed. But, what is more important, the heavy elements which were not present in the primordial gas also are shot out in large amounts

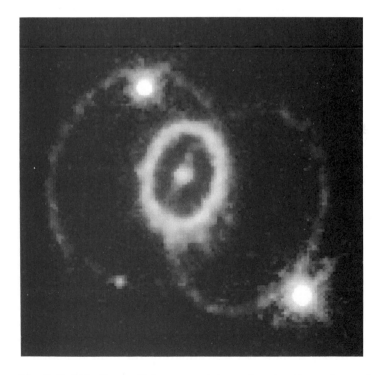

Fig. 5 Hubble Space Telescope picture showing three rings of glowing gas encircling the site of supernova SN 1987A in the nearby dwarf galaxy, the Large Magellanic Cloud. (Reproduced with permission of STScI and NASA.)

during the explosion. All this material, ejected at enormous velocities and expanding into interstellar space eventually becomes more or less mixed with the gas already there. Shock fronts within the volume of the growing supernova remnant cause gas locally to compress and in places eventually to collapse and form new stars out of this enriched interstellar material and these will have proportionally more of the heavier elements than the initial population of stars.

There are many other manifestations of violent mass loss from stars. A particularly intriguing case is the extreme, eruptively unstable star Eta Carinae, located about 9000 light years away in the southern sky. This object is one of the mysteries of stellar astronomy. It is one of the most massive and energetic stars in the Milky Way, about 100 times more massive and five million times more luminous than the Sun. In 1841 it flared up dramatically, becoming 600 times brighter than it is now, and subsequent observations showed that a compact nebula has started forming around the star. The best ground-based observations depict it as a small, fuzzy, elongated object. The Hubble Space Telescope image in Plate 7 reveals a remarkable wealth of detail within the nebula. A tenuous, red outer cloud surrounding the star is fast-moving material which was ejected from the interior of the star during the 1841 outburst. The inner, bright nebulosity also consists of ejected material but is denser and much dustier than the outer cloud. These two billowing lobes clearly have one moving towards us (lower left) and the other moving away. Bright jets emanate in the plane between the lobes. Eta Carinae has been observed several times with the Hubble Space Telescope and the nebula can be seen to be expanding faster near the centre than elsewhere within it. By tracing back the motions of hundreds of small-scale structures in the lobes it may be possible to shed light on how the lobes formed.

Returning now to the interstellar medium at large, the most obvious manifestations of interstellar material are the nebulae, seen by reflection, absorption, or self-emission. The planetary nebulae and supernova remnants which we have already met are examples of the latter. In reflection nebulae the dust grains preferentially reflect and scatter light from nearby stars. Dark nebulae, on the other hand, are dense clouds of gas and dust which have no suitable stars to illuminate them and are so opaque that they prevent light from background stars or bright nebulae from passing through them; they appear, in silhouette, as a 'hole' in the sky. Other types of nebulae not apparent on optical photographs are dense molecular clouds (containing mainly molecular hydrogen), which can be detected by their emission or absorption of radio, microwave or infrared radiation rather than of visible radiation. Spectacular emission nebulae, excited by hot stars, often are seen associated with these and actually are relatively small, hot cavities formed at the edges of or inside much more massive molecular clouds out of which the hot stars were formed.

A beautiful example of such a glowing cavity is the Orion Nebula in the constellation of Orion the Hunter. At a distance of about 1500 light years from the Sun it is visible to the naked eye as the hazy middle 'star' in Orion's Sword. Near the centre of the nebula is a small group of young, massive, hot stars which ionize and cause excitation of the nebula. The Hubble Space Telescope image in Plate 8 covers a typical small region about 1 light year across extending from the centre to the edge of the nebula. The tenuous, highly uneven distribution of gas in the nebula is clearly apparent, with the dust in the denser regions absorbing light so that gas lying in their shadows glows more dimly than surrounding regions. The nebula is one of the nearest regions of very recent star production (within the last million years) and in all contains several hundred stars at various early stages of formation.

When a star is born a great deal of cold gas and dust particles remain in a disk rotating around the star. Eventually protoplanets will form in regions of higher density within such disks and over time something approximating to the Solar System will evolve. Such protoplanetary disks were seen for the first time, and in abundance, in the Hubble Space Telescope images of the Orion Nebula. They are evident in Plate 6 as small concentrations of material ranging in shape from almost circular in appearance to variously elongated forms. Most appear bright against the background glow of the gaseous nebula, directly illuminated by the central bright group of stars; others, more compact, show as dark disks in silhouette. The large number of such objects in the Orion Nebula points to planet formation being a common occurrence in the Universe. Plate 9 is an enlargement of a region 0.14 light year across containing five young stars, four of which have evident surrounding disks comparable in size to the Solar System. The red glow at the centre of the dark disk is the newly formed star gleaming through the dense material.

Molecular clouds are the coldest form of matter in the Galaxy, with temperatures from a few degrees above absolute zero ranging to more than 100 degrees in regions of active star formation. The total amount of molecular gas in the interstellar medium probably exceeds the amount in atomic form. These clouds exist in a large variety of sizes, masses, and types. Most of the molecular gas in our Galaxy is in immense clouds of masses up to a million times that of the Sun, having diameters around 100 light years and central densities which may be more than 10 000 particles per cubic centimetre. These giant clouds are the largest self-gravitating bodies in the disk of the Galaxy and are the sites where most of the stars at present are being formed. The smallest clouds, the so-called Bok Globules, have masses ranging from only a few to some hundreds of Solar masses. Some of these have formed low mass stars and others may be collapsing on the way to forming stars. The Eagle Nebula is a star-forming region lying 7000 light years away in the direction of the constellation of Serpens. It contains many such dark globules and similar larger structures, contrasting against the bright emission nebula lit up by a loose cluster of hot stars. A Hubble Space Telescope image of the central region of the nebula (Fig. 6) shows astonishing pillar-like formations which protrude from the interior wall of the parent dark molecular cloud like stalagmites from the floor of a cavern; the tallest pillar is about a

Fig. 6 Hubble Space Telescope picture of part of M16, known as the Eagle Nebula. The dark pillar-like structures are slowly eroding, cool, dense gaseous regions that are incubators of new stars. (Reproduced with permission of STScI and NASA.)

light year in length. The pillars can be likened to buttes in a desert, where basalt or other dense rock has resisted erosion while the surrounding landscape has been worn away. In this celestial case the denser molecular cloud regions have survived longer than their surroundings in the intense flood of ultraviolet light from hot, massive, newborn stars which are off the top of the picture. The stars also illuminate the convoluted surface of the pillars, producing the dramatic visual effects which emphasize their three-dimensional nature. As the pillars themselves are slowly eroded away, small globules of even denser gas buried within them are uncovered. Many like these can be seen as finger-like protrusions near the top of Fig. 6. Each 'fingertip' is larger than the Solar System. Forming inside at least some of these are embryonic stars, which abruptly stop growing when they are separated from the parent reservoir of gas.

Taken together, molecular clouds and neutral atomic hydrogen clouds very far from fill the volume of interstellar space. Much of the gas between the clouds is very hot and very tenuous, at a temperature around a million degrees and density of a few particles per thousand cubic centimetres, being the direct outcome of the expansion of supernova remnants which have been produced in abundance throughout the Galaxy. We know that in a galaxy like ours a supernova explosion occurs once in every few tens of years and that the gaseous remnant will expand violently out to a radius of roughly 300 light years. Consequently, in a time as short as a few million years the gas at any point throughout the entire volume of the disk will be exposed at least once to a blast wave of hot gas from a supernova. The structure of the interstellar medium probably is dominated by a foam of such vast, hot, coalescing supernova remnants of various ages. The expanding front of a remnant progressing through the interstellar medium will envelop the denser clouds, and tend to sweep away any low-density material present and itself fill the space. The displaced material will be compressed into denser sheets or filaments in a shell around the remnants. After a time, this cools and then fragments to form an additional population of cold dense clouds. The various clouds move at an average speed of about ten kilometres a second; mutual collisions sometimes will lead to clouds coalescing, sometimes to the formation of several independent fragments. Clouds will also be disrupted by passing supernova remnants. Others will be reformed. As we have seen, new generations of stars arise, out of which in turn new supernovae erupt. This process of star formation, synthesis of heavy elements, expulsion of enriched material back into interstellar space, followed again by star formation, is a constant cycle of events which occurs throughout the Galaxy. The stuff out of which the Sun and the planets condensed, and out of which we are made, has been recycled many times in this way.

Galaxies

It is obvious from the appearance of other galaxies nearby that mostly they too have active interstellar media, although the gas content varies according to the type. Spiral galaxies (like the Milky Way) and irregular galaxies (like our low mass companion galaxies, the Large and Small Magellanic Clouds) contain the most gas, and star formation in them generally is very evident. Elliptical galaxies, on the other hand, the most massive as well as the most numerous of galaxies, contain much less gas and show fewer signs of star-forming activity. Elliptical galaxies are so named because of their smooth, overall ellipsoidal appearance (Fig. 7).

Galaxies are held in shape by a balance between the inward pull of gravity and the countering effect of stellar motions. In spiral galaxies stars move in nearly circular orbits confined to a disk structure (Fig. 7). In elliptical and irregular galaxies the stars swarm around in more random directions. How did these different forms come about? One idea is the following. Very early in the life of the Universe the violently expanding primordial 'soup' initially was almost but not quite smooth and featureless: there were small differences from place to place in the

Fig. 7 Ground-based telescope pictures of a typical elliptical galaxy (left) and a disk-like spiral galaxy seen face on (right).

density or expansion rate. Then, slightly overdense regions whose expansion rate lags behind the mean value evolve into huge, turbulent clouds whose internal expansion is eventually halted by self-gravitation. These protogalactic clouds continue contracting gravitationally while fragmenting into stars and eventually group into gravitationally bound clusters. The collapse can induce collisions between gaseous fragments which will lose energy and merge. With continual loss of energy the gas overall is increasingly less able to resist gravitation. A rapid collapse of the system ensues and, with a net rotational momentum in the original turbulent cloud, the end result will be a rotating disk flattened in the plane of rotation under its own gravitation. If stars are produced rapidly, however, before the gas has had time to settle down into a disk, since the dimensions of the stars are very small relative to the distances separating them there will be very little chance of collisions between them, so there will be no energy loss and they will retain their complex three-dimensional orbits. Whether there evolves a spiral or an elliptical galaxy therefore depends on whether the timescale for bulk star formation is longer or shorter than the collapse time of the protogalactic cloud. In an alternative scheme, at least some elliptical galaxies could form from later mergers of spiral galaxies in rich clusters.

One difficulty in understanding galaxies, however, is the fact that 90 per cent or more of their mass is unaccounted for. When we study the motions of stars and gas (by the Doppler effect) in the outer parts of galaxies we infer that the galaxies are experiencing the gravitational attraction of ten times more matter than we see. This 'dark' matter could be in one or more of several forms: faint very low mass stars, dead stars, or elementary particles of some as yet unknown type. To discover the character of such dark matter is one of the pressing issues of astronomy today.

After a galaxy has settled into a more or less stable dynamical structure it is still not unchanging thereafter. As we have seen earlier, a galaxy is above all an ensemble of stars which are born, evolve, and die—and therefore change luminosity and colour—and that progressively enrich the interstellar medium in chemical elements which they synthesize. Elliptical galaxies consequently are rather red in colour, typifying an ageing population of stars and the inability to form new stars in abundance because of the low interstellar gas content. Spiral galaxies are generally more luminous than the elliptical galaxies. They are rather blue, especially in the outer parts of the disk where much star formation is occurring, although a bulge in the centre where there is a concentration of older stars is more of a yellow-red colour, somewhat like the elliptical galaxies. Finally, the irregular galaxies contain a higher fraction of interstellar gas, with greater incidence of star formation, than do spiral galaxies and consequently have a dominant blue colour.

There is a class of galaxies which have 'peculiar' characteristics. Typically these show a bright, very concentrated, central region and an associated region of fast-moving hot gas not present in most galaxies, but otherwise appear as normal spiral or elliptical galaxies. The most extreme are the quasars where a region no bigger than the Solar System can outshine the entire surrounding galaxy a thousandfold. In such galaxies—known as active galaxies—a black hole as massive as 100 million Suns or more is understood to be the source of power, but there is no direct proof of this. A black hole like this can form when gas and stars accumulate in the centre of a galaxy until gravity overwhelms all other forces. The gravitational field is so strong that not even light can escape. Gas or even stars continue to fall into the centre and this captured debris furiously swirls downwards into the hole, moving close to the speed of light. Extreme frictional interactions in the spiralling matter can cause the conversion of at least 10 per cent of its mass to energy (from the equation $E = mc^2$) which can be radiated, giving it extraordinary luminosity.

Within the Virgo Cluster of galaxies lies one of the first known 'peculiar' galaxies, a giant elliptical galaxy 50 million light years away identified as M87, which is the brightest in the cluster. M87 has been long known to have a bright nucleus with an optical jet extending some 40 000 light years from it. It is also a radio source and an X-ray source. This galaxy has been much studied with the Hubble Space Telescope and has provided what most astronomers consider is conclusive evidence for a central black hole. The evidence is based on measurements of the rotational velocity of a swirling disk of hot gas surrounding the bright nucleus and being gravitationally influenced by the mass within it. On measuring the spectrum of emission lines from the hot gas at a close distance to the nucleus first on one side of the disk then on the other (Fig. 8) it was found—by the redshift and blueshift—that the gas is receding on one side and approaching on the other at the very high differential velocity of 1000 kilometres a second. The measurement of this orbital motion leads to a precise determination of the total mass contained within the small radius and this was found to be the amount of nearly three billion Suns. The small nuclear region is seen to contain only a tiny fraction of the number of stars required to make up this enormous mass. Consequently it is hard to avoid the conclusion that there is anything other than a supermassive black hole in the nucleus.

In contrast to M87 is the entirely normal galaxy M100, a member of the same cluster, which also has been imaged with the Hubble Space

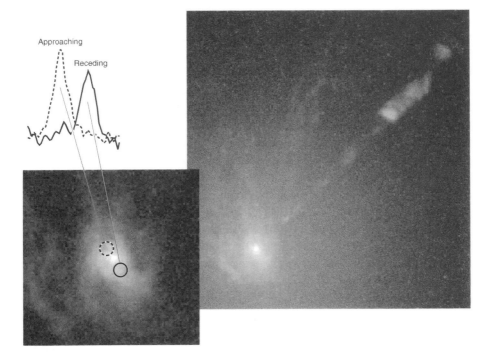

Fig. 8 Hubble Space Telescope pictures of the core of the peculiar giant elliptical galaxy M87 in the Virgo Cluster of galaxies, showing its jet and illustrating the spectroscopic measurement of the rotational velocity of the hot gas orbiting the centre. (Reproduced with permission of STScI, NASA, and Richard Sword).

Telescope (Plate 10). This majestic face-on spiral galaxy probably is closely similar to the Milky Way. Individual stars can be seen in the image as clearly as in galaxies only at a tenth of the distance viewed with ground-based telescopes. Distant though the galaxy is, the brightness variations of pulsating stars of a particular, rare, class, called Cepheid variable stars, could be measured in a sequence of images. These stars are known to be reliable distance indicators. A given star of this type shows a regular variation in brightness with a typical period of a few days, which is accurately related to its intrinsic brightness. Once the true brightness is thus obtained, the star's distance can be determined from its measured apparent brightness and application of the inverse square law of distance. From this measure of distance and of the recession velocity of the parent galaxy obtained from the redshift of lines in its spectrum, the constant of proportionality relating distance and velocity —known as the Hubble constant, after Edwin Hubble—is determined. This is one of the fundamental parameters of cosmology. It quantifies the present expansion rate of the Universe and allows a calculation of its age and development although other fundamental factors enter into this. Edwin Hubble used the same technique to obtain distances and this has been followed by many workers after him to improve the accuracy of the result, but the true value still is not known well. The importance of the Hubble Space Telescope measurements is the gain in accuracy achieved from the greatly increased baseline which can be used. Cepheid variable stars in many more galaxies will have to be measured, however, to ensure that local density perturbations leading to variations in velocity are fully recognized and allowed for. I believe a good value for the Hubble constant will come from such measurements within a very few years.

The enormous distances of galaxies makes it possible to study them in different stages of their evolution. The galaxies having the highest recession velocities are the most distant and therefore are those observed at earlier stages in their lives, according to the time when the light we see left them. The Hubble Space Telescope in this way enables us to examine galaxies at times when the Universe was only a few per cent of its present age.

A remarkable long-exposure image of a narrow region of the sky—in angular terms no larger than a large grain of sand appears at arm's length—which has been obtained by the Hubble Space Telescope is shown in Plate 11. This is our deepest, most detailed optical view of the Universe to date and shows galaxies which are so faint they have never before been seen even by the largest ground-based telescopes. Although this field is a very small sample of the heavens it is considered fully representative of the typical distribution of galaxies in space because the Universe, statistically, looks largely the same in all directions. The image can be likened to a core sample of the Earth's crust. In its long sightline the image contains objects progressively seen at many different stages in time; some date back to within only a billion years after the Big Bang. Enormous effort is going into the study of this and other similar images which have been obtained, but already several major differences from present day galaxy populations are clear. Whereas elliptical galaxies appear little different, many more spiral galaxies existed at early times than now, and generally these are in some way distorted or otherwise anomalous in appearance. There is also a large population of highly irregular or disturbed blue dwarf galaxies, many of them apparently interacting in pairs or small groups; the various forms of these include galaxies with superluminous star-forming regions, diffuse galaxies of various forms having very low surface brightness, and highly compact galaxies. Very few galaxies (or galaxy fragments) like these are seen in

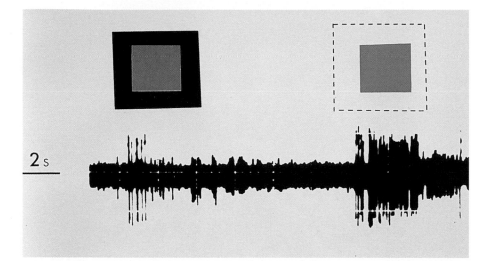

Plate 1 The responses of a cell in area V4. This cell responds best to a blue square on a white background.

Plate 2 *Red Square* by Kazemir Malevich.

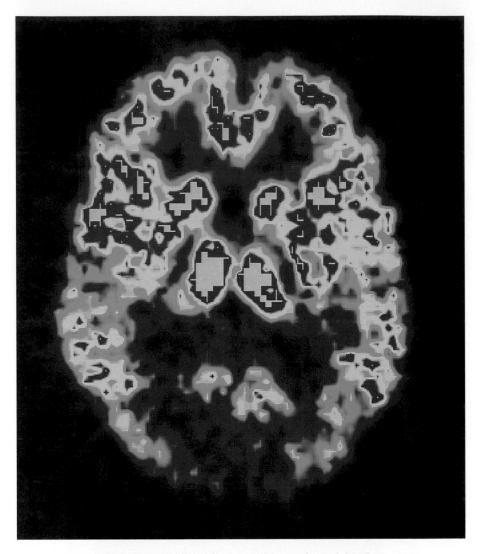

Plate 3 PET scan for [^{11}C]diprenorphine, a ligand for the opiate system, showing uptake at specific receptor sites in the brain. (From a photograph kindly provided by the MRC Clinical Sciences Centre.)

Plate 4 Successive Hubble Space Telescope pictures of Jupiter showing the influence of winds on some Comet Shoemaker–Levy 9 impact sites over a time of 5 days. (Reproduced with permission of the Space Telescope Science Institute (STScI) and NASA.)

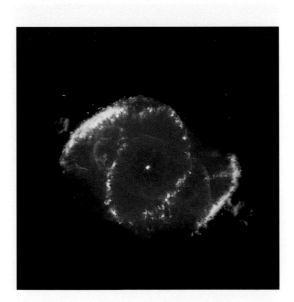

Plate 5 Hubble Space Telescope picture of the planetary nebula NGC 6543, nicknamed the Cat's Eye Nebula. (Reproduced with permission of STScI and NASA.)

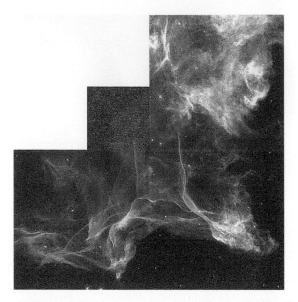

Plate 6 Region of the supernova remnant known as the Cygnus Loop imaged with the Hubble Space Telescope. The stepped outline of this and other pictures in this article comes from the way the images are assembled. (Reproduced with permission of STScI and NASA.)

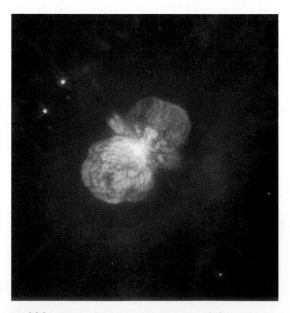

Plate 7 Hubble Space Telescope picture of the supermassive, eruptively unstable star Eta Carinae showing billowing lobes of gas and dust. (Reproduced with permission of STScI and NASA.)

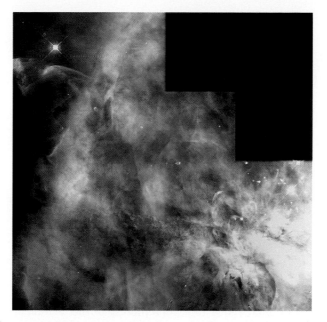

Plate 8 Hubble Space Telescope picture of a region in the Orion Nebula. (Reproduced with permission of STScI and NASA.)

Plate 9 Enlarged view of a region in Plate 8 containing young stars with protoplanetary disks of gas and dust. (Reproduced with permission of STScI and NASA.)

Plate 10 Hubble Space Telescope picture of the face-on spiral galaxy M100 in the Virgo Cluster of galaxies. (Reproduced with permission of STScI and NASA.)

Plate 11 The deepest, most detailed optical view of the Universe yet obtained, showing galaxies in the early Universe. This Hubble Space Telescope picture covers a speck of sky only one-thirtieth of the diameter of the full Moon. (Reproduced with permission of STScI and NASA.)

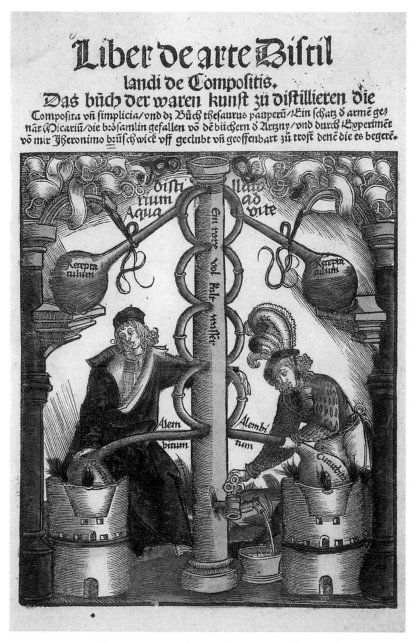

Plate 12 An early distillation apparatus

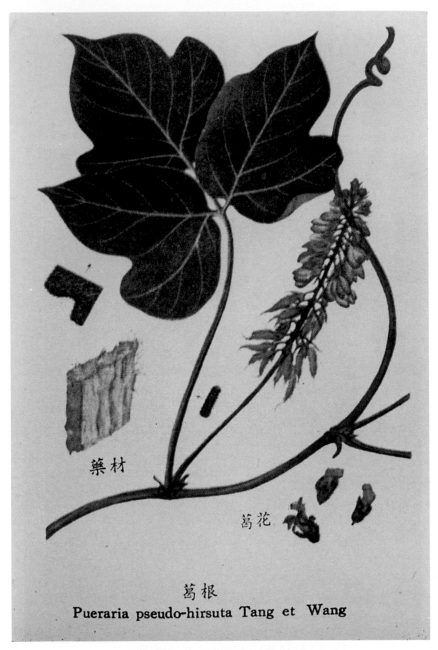

藥材

萏花

萏根
Pueraria pseudo-hirsuta Tang et Wang

Plate 13 The kudzu plant

the Universe today. The observations confirm that elliptical galaxies developed remarkably quickly into their present shapes, whereas spirals evolved over a much longer period. It seems that the majority of spiral galaxies were first built and then torn apart during interactions and collisions. Galactic 'cannibalism' also was far more common in the early Universe: those galaxies like the Milky Way which have survived probably grew to their current sizes by consuming their smaller neighbours.

Gravitational lenses

Apparent giant arcs seen in the direction of some rich clusters of galaxies are known to be the magnified, brightened, and distorted shapes of distant background galaxies. These curious images are produced by the gravitational field of the massive cluster which deflects the light as it passes through it—much as an optical system bends light—causing it to behave as an enormous gravitational lens. This provides a powerful 'zoom lens' for observing detail in galaxies that are too distant to be seen clearly if at all. Such arcs are rather rare phenomena which occur only when the background galaxy is viewed in a rather close alignment with the mass centre of the dense foreground cluster. The presence of such arcs demonstrates that the total mass in galaxy clusters is much greater than can be accounted for from the galactic luminous matter alone, confirming that clusters contain large amounts of dark matter, as has been deduced from the dynamical behaviour of stars and gas in galaxies as well as of galaxies in clusters. From a study of these arcs and indeed of the weaker deformation of the images of the entire population of background galaxies, this lensing process gives crucial information both on the mass distribution in the lensing cluster and on the nature of distant galaxies.

The Hubble Space Telescope has produced startling images of galaxy clusters exhibiting such giant arcs. One of these, the rich cluster Abell 2218, is shown in Fig. 9. The numerous arcs in the cobweb-like pattern spread across the picture are difficult to detect with ground-based telescopes because they are so thin. A striking result possible only because of the Hubble Space Telescope's fine imaging capability was the detection of internal structure in these slender arcs showing that the separate images are all of the same background source. Through a computational 'de-lensing' process, using the measured mass distribution of the cluster, the true shape of the distant galaxy can be constructed. This combination of natural optics in the sky with the Hubble Space Telescope's wonderful imaging capability opens tremendous possibilities for probing early cosmic times.

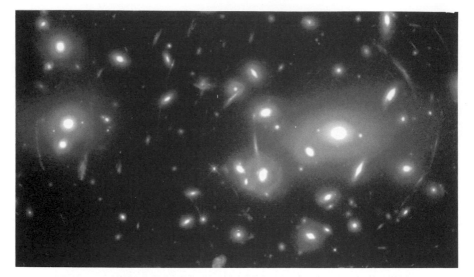

Fig. 9 Hubble Space Telescope picture of the massive galaxy cluster Abell 2218. The arc-like patterns arise from the presence of a galaxy in the distant background, which is gravitationally imaged by the cluster. (Reproduced with permission of STScI and NASA.)

Concluding remarks

The Hubble Space Telescope has proved a magnificent scientific research tool, exceeding even the highest expectations. Astronomers are receiving pictures of the heavens which are as perfect as they can be. Important new discoveries always follow when major new instruments or techniques appear and this is doubly true in astronomy because of the enormously wide range of applications to which an instrument is put by the army of successful bidders for precious observing time. The competition is so fierce that only the top 20 per cent or so of the proposals can be awarded time; because of this evident competition only proposals in the 'excellent' class are submitted. This heady combination of instrumental and scientific excellence will ensure that the Hubble Space Telescope continues in its now indispensable role for a long time to come. Undoubtedly it will provide an accurate measure of the expansion of the Universe. Most probably it will show how and when the first galaxies formed. Certainly it will add invaluably to the fabric of our knowledge in topics across the whole of astronomy. However, it must not be forgotten that ground-based telescopes are equally indispensable. As just one example, only with the large ground-based telescopes, with their enormous light grasp, are we able to examine spectroscopically and understand the faintest objects found with the Hubble Space Telescope. But that is for another occasion.

ALEC BOKSENBERG

Born 1936, graduated and obtained his doctorate from the University of London. Held positions leading to Head of the Ultraviolet and Optical Astronomy Research Group and Professor of Physics and Astronomy at University College, London. There he developed new instrumentation for space and ground-based astronomy and made studies which established a new field of research in observational cosmology. In 1981 he moved to the Royal Greenwich Observatory as Director and was responsible for establishing Britain's three major optical telescopes, the Isaac Newton Group, on the island of La Palma in the Canary Islands. In 1993 became overall Director of the Royal Observatories, then including the Royal Observatory, Edinburgh, with its infrared and millimetre-wave telescopes in Hawaii. Holds a Visiting Professorship of University College, London, and Honorary Professorship of the University of Cambridge, and has been awarded several honorary doctorates. He is the author of more than 200 research papers on subjects in astronomy. Since 1996 he is PPARC Research Professor in the University of Cambridge. Has been associated with the Hubble Space Telescope since 1972, in particular with the European Space Agency's instrument, the Faint Object Camera, now operating on board the HST and whose design is centred on the ultra-sensitive Image Photon Counting System detector which he invented. In 1996 he was made CBE.

Sheathing the two-edged sword—one hundred years of radioactivity

PETER CHRISTMAS

Introduction

It is appropriate that a review of the discovery of radioactivity one hundred years ago and a discussion of the consequences of that discovery for science and the world at large should be undertaken at the Royal Institution.

The discovery of radioactivity by Henri Becquerel owes much to the researches of Michael Faraday. Faraday came to the Royal Institution in 1812, at the age of 21, to work with Sir Humphrey Davy. Over the next 46 years his research covered a wide range of topics but the most important aspect, both for science at large and for the discovery of radioactivity, was his work on electricity and electromagnetism.

Of particular relevance was Faraday's work on electrolysis and electrical discharges in gases, subjects which received much attention in the latter half of the nineteenth century. By the mid-1890s it had already been observed that cathode rays, produced by an electrical discharge in a gas at low pressure, caused fluorescence when they hit the wall of the glass vessel in which they were produced; in 1897 John Joseph Thompson showed that the cathode rays were identical to electrons, the name proposed by G.J. Stoney in 1894 for the particles carrying the fundamental unit of negative charge in electrolysis, identical in magnitude to the positive charge carried by the hydrogen ion.

In 1895, in the course of further studies of electrical discharges in gases, X rays were discovered by Wilhelm Conrad Röntgen working at the University of Wurtzburg. Röntgen observed that a paper screen, covered with barium platinocyanide and placed near to his discharge tube, fluoresced when the apparatus was switched on, even though the latter was covered by opaque paper and the room was in darkness. In December

1895 Röntgen published his findings under the title *On a New Kind of Ray*, for which he coined the name 'X rays'. The most notable property of the X rays was their ability to penetrate materials opaque to visible light, a property which was immediately seized upon and exploited with an enthusiasm typical of the age. In 1901 Röntgen received the first Nobel Prize for Physics for his discovery of X rays.

X-ray pictures are not produced by 'reflected light' in the same way as ordinary photographs; they are records on film of shadows cast by objects which are denser than the surrounding medium, just as shadows can be formed with visible light; the eye is of course insensitive to X rays. These elementary facts were not, however, widely understood. Thus in 1896, the *London Pall Mall Gazette* claimed already to be 'sick of the Röntgen rays ... you can see other people's bones with the naked eye, and also see through eight inches of solid wood. On the revolting indecency of this there is no need to dwell'. An enterprising London firm even advertised 'X-ray proof underclothing for ladies'.

Following the discovery of X rays the search immediately began for other sources of penetrating radiation. One seemingly promising line of enquiry was based upon the observation that X rays appeared to be associated with the strong fluorescence produced by cathode rays.

Of all the people engaged in this search, none was better placed to succeed than the Frenchman Henri Becquerel, by 1896 an accomplished and respected physicist and member of the French Academy of Sciences. Becquerel was born in 1852, into a prominent scientific family. His early researches included the study of the rotation of plane-polarized light by magnetic fields, an effect discovered by Michael Faraday in 1845 and named after him, providing another link, albeit tenuous, between radioactivity and the Royal Institution.

In 1892 Becquerel had become professor of applied physics at the National Museum of Natural History, a post held by his father and grandfather before him, and later to be held by his son (see Fig. 1). The Museum was in fact an active research centre, and the chair in physics seems to have been created specifically for Henri Becquerel's grandfather Antoine-César Becquerel. Be that as it may, this particular post was held continuously by the Becquerel family from 1838 until 1953, when Jean Becquerel died.

Henri Becquerel was an accomplished experimentalist, with expertise in phosphorescence and a familiarity with uranium compounds. Early in 1896, working in his laboratory at the family home (*Maison Cuvier, Jardin des Plantes*), Becquerel set out upon a systematic search for penetrating radiations from fluorescent materials, using as his detector photographic plates wrapped in opaque paper to exclude visible light.

Fig. 1 Henri Becquerel, on the left, with his father, Antoine and son, Jean. The photograph dates from about 1882. (Reproduced courtesy of H. Zurfluh, Musée de Chatillon-Coligny.)

On 24 February 1896 he reported to the Academy of Sciences that he had observed penetrating radiation, similar to X rays, from uranium potassium sulfate which had been placed upon his detector and made to fluoresce by exposure to sunlight.

On Sunday 1 March, for reasons which remain obscure to this day, Becquerel developed some plates which had been coated with the uranium salt some days before but never exposed to sunlight owing to bad weather; the uranium salt had not fluoresced, but had been in contact with the photographic plate for longer than usual. Expecting to find very weak images, he was surprised to find an even more pronounced blackening. Besides clear images (autoradiographs) of the individual uranium salt crystals, there was also the shadow of a small copper cross which Becquerel had placed between one of the crystals and the photographic plate.

Becquerel announced his discovery of penetrating radiation from uranium *sans fluorescence* at the Academy meeting on 2 March 1896. Interestingly, an Englishman, Sylvanus Thompson, in London, had discovered a similar effect with uranium nitrate at about the same time, but Becquerel was the first to publish his findings.

In subsequent experiments Becquerel showed that the penetrating rays were produced by all uranium salts, regardless of whether or not they could be induced to fluoresce, and by uranium metal. The emissions appeared constant with time (due, as later emerged, to the long half life of uranium), they were unaffected by any physical or chemical agent, and, like X rays, they caused the discharge of electrified bodies. Becquerel had clearly discovered an entirely new phenomenon, associated with the element uranium: the name 'radioactivity' was proposed by Marie Curie, who showed that it was a property of the uranium atoms (see below).

Marya Salome Sklodowska was born and grew up in Warsaw. She went to Paris in 1891, at the age of 23, and studied physics and chemistry, followed by mathematics, at the Sorbonne. In 1894 she met the French physicist Pierre Curie and they were married on 26 July 1895. Pierre Curie had already established his reputation as a physicist; with his brother he had discovered piezo-electricity, and in magnetism the Curie point and Curie's law are both named after him. Pierre Curie now joined his wife to work with her on radioactivity, a collaboration which continued until Pierre's untimely death in a street accident in 1906.

Polonium and radium

The discovery of radioactivity attracted, initially, far less attention than that of X rays, largely due to the much greater intensity with which the latter could be produced, as compared with the low intensities of penetrating rays emitted by uranium. However, Becquerel, the Curies, and others continued to study the new phenomenon.

The earliest work on radioactivity depended, as we have seen, upon photographic detection, but Becquerel was quick to exploit his discovery that the rays from uranium, like X rays, could cause an electric current to flow between charged metal plates within a gas-filled volume. This effect was studied in Cambridge by J.J. Thompson, assisted by a young research fellow from New Zealand named Ernest Rutherford, and was shown to be due to the ability of the radiation to split the electrically

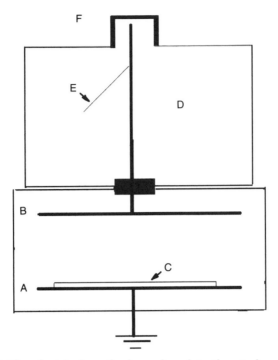

Fig. 2 The electrical method employed in the study of radio-activity. Further details may be found in the text.

neutral gas atoms into negatively charged electrons and positively charged atoms or ions. This process is called ionization, hence the term 'ionizing radiation'.

Figure 2 shows the arrangement used by Becquerel. The equipment comprised an ionization chamber connected to a gold-leaf electroscope. The electroscope was charged by removing the insulating cap, F, and applying a voltage, causing the gold leaf, E, to rise as shown. The uranium compound, C, was then placed upon the lower plate, A, of the ionization chamber, whereupon an ionization current was produced, causing the gold leaf to return to its original position. The rate at which the gold leaf moved was proportional to the amount of radioactivity present.

This type of equipment was capable with care of detecting electrical currents as small as 10^{-15} A; it was far more sensitive than the photo-graphic method and became widely used in the search for other radio-active materials.

In 1898 G.C. Schmidt and, independently, Marie Curie, showed that thorium and its compounds were also radioactive. In a crucial series of experiments, Marie Curie used the electrical method to compare the radioactivity of a large number of synthetic compounds and naturally occurring minerals containing uranium and thorium. She found that the activity of synthesized uranium compounds was always proportional to the elemental uranium content, from which it was deduced that radio-activity is an atomic property. She observed, however, that the activities of some naturally occurring minerals, such as pitchblende (uranium oxide) or chalcolite (hydrated uranyl copper phosphate), were several times higher than was to be expected from the amounts of uranium or thorium present.

Marie Curie concluded that the higher activity of some minerals was due to the presence of an unknown substance or substances with much higher activity than uranium or thorium. In a masterly series of chemical analyses the Curies then proceeded to isolate two new active substances; that which separated with bismuth they named polonium, after Marie Curie's country of birth, and that associated with barium they called radium. Pure radium chloride was then separated from barium by fractional crystallization.

In due course the Austrian government provided a ton of uranium residues from the Joachimsthal mines in Bohemia, from which Marie Curie extracted 100 mg of pure radium salt. She then used this material to determine the atomic weight of radium, obtaining the value 225, close to the modern value of 226. This was chemistry on an heroic scale, the processing of a ton of ore requiring five tons of assorted chemicals and fifty tons of water.

Properties of the penetrating radiations

Hand-in-hand with the search for new radioactive materials went the quest for a greater understanding of the nature of the emitted radiations and the processes giving rise to them.

In 1898 Ernest Rutherford left Cambridge to became Professor of Physics at McGill University in Montreal. As with many a young person both before and since, this was a career move motivated by the wish to improve his financial position in order that he might marry; he wrote to his fiancée Mary Newton, 'the salary is only £500 but enough for you and me to start on'.

In 1899 Rutherford reported, from a study of their penetrating power, that the rays from uranium comprised a less penetrating component which he called 'α rays' and a more penetrating form which he termed 'β rays': α rays could be stopped by a sheet of paper, while β rays could pass through several millimetres of aluminium.

Taking advantage of the much greater activity of radium as compared with uranium, Paul Villard showed in 1900 that a third component of radiation was also emitted, similar to X rays in not being deflected by a magnetic field, but more penetrating than any of the previously discovered emissions, a result confirmed by Becquerel. These exceptionally penetrating rays, requiring several centimetres of lead to stop them completely, were called 'γ rays'; they were first observed by covering a radium source with a few millimetres of lead to absorb all the α and β rays. In due course the γ rays were shown to be diffracted by crystals in the same way as X rays, and their wavelengths were determined. X rays and γ rays are both forms of electromagnetic radiation. The penetrating properties of α, β, and γ rays can be demonstrated very conveniently using sources of ^{241}Am, ^{36}Cl, and ^{133}Ba respectively.

In 1899 Friedrich Geisel showed that β rays were deflected by a magnetic field in the same direction and to the same extent as cathode rays, and soon after this Becquerel showed from their deflection in magnetic and electric fields, that the ratio of charge to mass, e/m, for the β rays was close to that of J.J. Thompson's electron: it became clear that β rays were indeed electrons. In 1903 Rutherford showed that α rays were also deflected by electric and magnetic fields, though to a several hundred times smaller extent than the β rays. Subsequent measurements of the deflection of α rays in electric and magnetic fields lent credence to the hypothesis that they were helium atoms with a positive charge equal in magnitude to twice the charge of the electron, and this was finally confirmed in 1909 by Rutherford and his student Thomas Royds by a direct experiment. Since that time the terms 'particle' and 'ray' have both been used in referring to these radiations.

It was also established that α rays were emitted with well-defined energies whereas the β-ray energies were variable. This puzzle was not solved until 1931 when Wolfgang Pauli postulated that each emitted β ray is accompanied by a second particle, called a neutrino (ν), which shares the total energy available. The neutrino carries no electric charge and has a very small, possibly zero, mass; it is thus very difficult to observe and its existence was not confirmed until 1954.

The particulate nature of the α rays was nicely demonstrated by the 'spinthariscope', invented by Sir William Crookes in 1903, in which the impacts of individual α particles on a zinc sulfide screen are observed as tiny scintillating points of light. This effect was quickly put to use as a method of measuring radioactivity by counting the individual scintillations, providing an early example of the advantage of digital as opposed to analogue methods.

Recognition

In 1903 the Nobel Prize for Physics was awarded jointly to Henri Becquerel for his discovery of radioactivity and to Pierre and Marie Curie for their work on radiation phenomena. Becquerel's name was incorrectly spelt on his Nobel Prize certificate; fortunately this did not invalidate the award!

Henri Becquerel gave a discourse at the Royal Institution in 1902, followed by Pierre Curie in 1903: it was recorded that the Curies were received enthusiastically; they were later awarded the Davy medal of the Royal Society. By this time the remarkable properties of radium were well known and the Curies had become world-famous, as indicated, for example, by a cartoon, Fig. 3, which appeared in the magazine *Vanity Fair* in 1904.

Fig. 3 A cartoon depicting Pierre and Marie Curie, published in *Vanity Fair* in 1904 with the caption 'Radium'. (Reproduced by kind permission of the Wellcome Institute Library, London.)

Sadly, Henri Becquerel died suddenly from a heart attack in 1908, at the early age of 55, and so did not live to see many of the consequences of his discovery.

In 1911 Marie Curie received further recognition in the form of the Nobel Prize for Chemistry, for the discovery of radium. Chemists and physicists continue to work closely together in radioactivity, particularly in the field of measurements and standards.

Transformation theory

As has so often been the case in other branches of science, the newly discovered phenomenon of radioactivity found applications long before it was properly understood. In the early work on uranium, thorium, radium, and polonium, it had been assumed that the property of radioactivity was permanent, showing no decrease with time. However, in 1900, Rutherford showed that thorium emitted a radioactive gas, or 'emanation' which lost half its activity in less than one minute. In that same year, Rutherford was joined by a young English chemist called Frederick Soddy, beginning a partnership which was to play a major role in unravelling the mysteries of radioactivity.

In 1902 Rutherford and Soddy announced the discovery of thorium X, a short-lived and chemically distinct product of thorium, analogous to uranium X, discovered in 1900 by Sir William Crookes; both were short-lived. The decay of these short-lived species is characterized by the fact that the radioactivity always decreases by a factor of two in a characteristic time, called the 'half life'.

In 1903, on the basis of these discoveries, Rutherford and Soddy proposed their so-called 'transformation theory', according to which the radioactive elements are unstable and undergo spontaneous disintegration, or decay, to give substances with distinct chemical properties which may themselves, in turn, be radioactive. The α, β, and γ rays carry off most of the excess energy which is associated with the instability.

Uranium was shown to be the starting point of a series of successive transformations, which includes radium and radon and ends with a stable form of lead; a similar series was identified for thorium, and for actinium, which had been discovered by André Debierne in 1899. Radioactivity appeared to be restricted to the heavy elements, apart from potassium and rubidium, both of which were observed to be weakly radioactive.

The characteristic half life arises because radioactive decay is exponential, the rate at which atoms transform being strictly proportional to

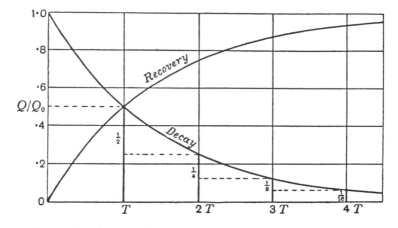

Fig. 4 The decay and recovery curves of radon; see text for details. (Reproduced from E. Rutherford *et al.*, 1930, by kind permission of Cambridge University Press.)

the number of undecayed atoms which are present. The curve labelled 'decay' in Fig. 4 shows the exponential decay of a freshly separated radon sample. For radon the half life, T, is about 3.8 days; radon is produced by the decay of radium, for which the half life is very much longer, some 1600 years. The curve marked 'recovery' shows how, in a newly produced radium sample, the radon activity, initially zero, builds up until, after a time interval approaching $10 \times T$, the radon is decaying as fast as it is produced: the radon and radium are then said to be in radioactive equilibrium.

Rutherford was evidently rather captivated by the notion of exponential decay; when he was created Baron Rutherford of Nelson in 1931, he incorporated a pair of exponential curves into his new coat of arms!

The birth of nuclear physics

In 1907 Rutherford returned from Canada to England to take up a chair in physics at Manchester University. Among his co-workers at Manchester over the next few years were Hans Geiger and Ernest Marsden, both of whom were to make major contributions to the developing understanding of radioactivity.

In 1908 the Manchester group embarked upon a systematic study of the scattering of α and β rays. Early experiments, in which beams of these radiations were passed through thin metallic foils, showed that the α particles were much less scattered than were β rays, as was to be expected from their much greater kinetic energy, and Geiger showed that the average angle of scatter increased with the atomic weight of the foil. The results observed thus far could be explained in terms of a probability distribution based on the measured average scattering angle.

However, in 1909 the crucial discovery was made, by Geiger and Marsden, that occasionally an α ray was scattered through such a large angle that it re-emerged from the front face of the scattering foil. The importance of this discovery lay in the insight which it provided for Rutherford about the internal structure of atoms; it led directly to the modern theory of atomic structure.

The theory that all matter is composed of atoms had been proposed in its modern form by the English chemist John Dalton in 1803 and had become well established as chemistry developed into a quantitative subject. Dalton believed that atoms were indivisible, but it gradually became clear that this might not be so. Important clues came from Dmitri Ivanovich Mendeleev, who drew attention to the regular occurrence of similarities between the elements.

In 1869 Mendeleev produced his 'Periodic Table of the Elements', Fig. 5. On the basis of the systematics underlying this table, Mendeleev correctly predicted the properties of the then aˢ yet undiscovered elements scandium, gallium, and germanium, needed to fill the three gaps in the middle of the table. Around the turn of the century the column on the left was filled by the discovery of the noble gases helium, neon, argon, krypton and xenon, and the gaps at the bottom were now to be filled by the radioactive elements. The two remaining gaps in the 1869 table were filled by the discoveries of hafnium in 1923 and the wholly radioactive

O	I	II	III	IV	V	VI	VII	VIII
							H	
	Li	Be	B	C	N	O	F	
	Na	Mg	Al	Si	P	S	Cl	
	K	Ca		Ti	V	Cr	Mn	Fe Co Ni
	Cu	Zn			As	Se	Br	
	Rb	Sr	Y	Zr	Nb	Mo		Ru Rh Pd
	Ag	Cd	In	Sn	Sb	Te	I	
	Cs	Ba	Rare Earths		Ta	W	Re	Os Ir Pt
	Au	Hg	Tl	Pb	Bi	Po	As	
				Th		U		

Fig. 5 Mendeleev's 'Periodic Table of the Elements', 1869.

element technicium in 1937. More recently, the table has been enlarged by the discovery of artificially produced heavy elements (see below).

It was becoming clear that there must be some internal structure to atoms, giving rise to the systematic behaviour observed by Mendeleev. At last, physicists became interested in atoms, clearly regarding them as too important to be entrusted solely to chemists!

In 1903 J.J. Thompson had proposed his 'plum pudding' model of the atom, with negative electrons embedded in a sphere of positive electric charge: this model might have explained ionization, but it could not account for the α-particle scattering observed by Geiger and Marsden, an effect described by Rutherford as about as likely as the deflection of a 15-inch cannon shell by a sheet of tissue paper!

In 1911, however, Rutherford showed that the scattering of the positively charged α particles observed by Geiger and Marsden could be explained if all the positive charge of the target atom, and most of its mass, were concentrated into a very small central 'nucleus'. The electrons were perceived as moving in circular orbits around the nucleus, just as the planets move in orbits around the sun.

If a typical Rutherford atom were enlarged until it filled the lecture theatre at the Royal Institution, the nucleus would be of order 1 mm in diameter. If the nucleus were as large as Rutherford's cannon shell, then the atom would correspond to a sphere enclosing London (Peter Kalmus has pointed out that a sheet of tissue paper the size of London would be several thousand times heavier than the shell; concentrated into a comparable volume this mass would be more than sufficient to reflect the shell in the event of a direct hit!).

Despite its success, there was a major difficulty with the 'planetary' model of the atom, as Rutherford himself acknowledged in 1913: according to the known laws of physics, the orbiting electrons should lose energy and collapse into the nucleus. The solution was provided by the Danish physicist Neils Bohr, who had become interested in the problem while working with Rutherford in Manchester for four months in the spring of 1912. Bohr made use of the newly discovered quantum theory.

For the origins of quantum theory we need to return once again to the work of Michael Faraday. Faraday's work on electricity led him to the notion of the electromagnetic field with electric and magnetic lines of force and to a connection between optics and magnetism. These ideas were developed by James Clerk Maxwell in the 1860s. Maxwell showed mathematically that an oscillating electrical charge generates a transverse electromagnetic wave which propagates with the velocity of light. This prediction was verified experimentally by Heinrich Hertz, who built an oscillating electrical circuit to generate electromagnetic waves which could be reflected, refracted and polarized just like visible light.

Despite the success of Maxwell's theory, some serious difficulties soon arose. In particular, the theory failed to explain the observed distribution of radiation from a very hot, or incandescent body. According to the classical theory, the intensity should become infinitely large at high frequencies, rather than reducing to zero as observed. This prediction was called the 'ultraviolet catastrophe'; it is interesting to note, in the present context, that the region for which the classical theory failed included the newly discovered X and γ rays!

A means of avoiding the ultraviolet catastrophe was provided by Max Planck in 1900. Planck postulated that, at the atomic and subatomic level, emission and absorption of radiation do not take place continuously, but only in finite amounts, or 'quanta', of value $h\nu$ where ν is the frequency and h is a universal constant, now called Planck's constant. On this basis Planck was able to derive an equation which exactly reproduced the observed variation of thermal radiation intensity with frequency.

The quantum theory was not readily accepted at first, but was made more respectable by Albert Einstein, who showed that it could provide answers to other problems which defied solution in terms of classical physics, such as the variation of specific heat with temperature. Einstein went further than Planck and postulated that light itself is quantized, such that light of frequency ν consists of discrete particles, now called photons, of energy $h\nu$. In this way he was able to explain the observed photoelectric effect, in which light falling on a metal surface in a vacuum ejects electrons whose energy depends only upon the frequency of the light, and not on its intensity.

Neils Bohr took the ideas of quantum theory and applied them with brilliant success to Rutherford's nuclear atom. He assumed that atoms could exist only in discrete stationary states which could not emit or absorb radiation in arbitrarily small amounts. Such emission or absorption, of photons of energy $h\nu$, could occur only in conjunction with a transition from one stationary state to another with energy difference $h\nu$. On this basis Bohr was able to explain not only the stability of the planetary atom but also the optical line spectrum from excited hydrogen atoms: He showed that the observed lines, comprising the so-called Balmer series, correspond to transitions between calculable discrete energy levels of the single electron in the neutral hydrogen atom.

With the development of wave mechanics and the notion of the electron as having intrinsic angular momentum, like a spinning top, the Bohr model was expanded in due course to account also for the periodic

CARL A. RUDISILL LIBRARY
LENOIR-RHYNE COLLEGE

properties of the elements as observed by Mendeleev and for observed X-ray, as well as optical, spectra.

From isotopes to artificial radionuclides

In December 1913, in the journal *Nature*, in a two-page paper entitled '*Intra-atomic Charge*', whose brevity belied its importance, Frederick Soddy introduced the notion of 'isotopes' and discussed further the nature of the planetary model. By this time a large body of work, by Soddy and others, had shown that there were far too many apparently distinct new radioactive elements to fill the available gaps in Mendeleev's periodic table.

Soddy noted, however, that a substantial number of these new species were chemically indistinguishable, differing only in their radioactivity and their atomic weight, and he called them 'isotopes' to indicate that they should occupy the same place in the periodic table; work on the homogeneity of neon, also published in 1913, gave support to the notion that all elements could exist as isotopes. On this basis the known elements fitted neatly into the periodic table, from the lightest, hydrogen, to the heaviest then known, uranium.

By this time the nucleus had come to be widely regarded as composed of a sufficient number, M, of positively charged hydrogen atoms to provide the nuclear mass, together with about half this number of electrons, to give the correct nuclear charge, Z, or atomic number, defining the position of the element in the periodic table. This view of the atom was not greatly dissimilar to Prout's hypothesis of some 80 years earlier, that all matter is composed of hydrogen atoms.

In 1920, in his Bakerian Lecture to the Royal Society, Rutherford, by then Sir Ernest (he was knighted in 1914 for his services to science) proposed the name 'proton' for the positively charged component of the nucleus and speculated that the other component of the nucleus might be a neutral particle formed in some way from a proton and an electron. The true nature of this neutral particle was revealed, as we shall see, in 1932.

The introduction of isotopes led to a reduction in the number of elements involved in the heavy element decay series; in the case of the complete uranium series (Fig. 6), the number of participating elements was reduced from 18 to 10.

We have seen how the subject of nuclear physics developed from the work on α-ray scattering by Rutherford and his collaborators at Manchester. In 1919 Rutherford made the further important discovery

URANIUM I	URANIUM - 238
↓	
URANIUM X₁	THORIUM - 234
↓	
URANIUM X₂	PROTOACTINIUM - 234
↓	
URANIUM II	URANIUM - 234
↓	
IONIUM	THORIUM - 230
↓	
RADIUM	RADIUM - 226
↓	
RADON	RADON - 222
↓	
RADIUM A →	POLONIUM - 218
↓	
RADIUM B ↓	LEAD - 214
↓ ASTATINE	ASTATINE - 218
↙	
RADIUM C →	BISMUTH - 214
↓	
RADIUM C' ↓	POLONIUM - 214
↓ RADIUM C"	THALLIUM - 210
↙	
RADIUM D	LEAD - 210
↓	
RADIUM E →	BISMUTH - 210
↓	
RADIUM F ↓	POLONIUM - 210
↓ THALLIUM	THALLIUM - 206
↙	
RADIUM G	LEAD - 206
(stable)	

Fig. 6 The uranium decay series.

that α particles could themselves cause nuclear transformations to take place, a process described as 'nuclear chemistry'. In the first example, studied by Rutherford using the cloud chamber invented by C.T.R. Wilson in 1897, nitrogen gas bombarded with α particles was transformed into oxygen with the simultaneous production of an energetic proton; using the notation $^{M}N_{Z}$ this process is written as

$$^{14}N_7 + {}^4He_2 \rightarrow {}^{17}O_8 + {}^1H_1.$$

Subsequently, many other examples of nuclear chemical reactions were discovered.

In 1932 Cockroft and Walton built the first particle accelerator, employing high voltages to accelerate positively charged hydrogen atoms, i.e. protons, to energies sufficiently high for them to be used as bombarding projectiles in the same manner as α particles. It was found that certain of the lighter elements, such as lithium and beryllium, when bombarded with protons in this way, gave out a new form of very penetrating radiation which could in turn cause protons to be ejected from materials containing hydrogen.

In 1932 J. Chadwick showed that the new radiation comprised an uncharged particle of mass approximately equal to that of the proton which he identified with Rutherford's proposed neutral constituent of the nucleus and which he called a 'neutron'. It became clear that the neutron is a particle in its own right, and not a combination of a proton and an electron. Neutrons and protons, known collectively as nucleons, are the bricks from which all atomic nuclei are built; understanding the forces which hold these bricks together, and their consequences, is the stuff of nuclear physics.

In 1934, Marie Curie's daughter Irene Joliot-Curie and her husband Frederick showed that when α particles from her parents' polonium were used to bombard light elements the products were sometimes radioactive. Enrico Fermi and his collaborators showed that similar results could be obtained with neutrons. These discoveries of artificially produced radionuclides marked a turning point in the applications of radioactivity, up to now largely confined to radium and radon.

It was found also that some of the new, artificially produced radionuclides undergo a form of β decay in which a positively charged electron or 'positron' is emitted. This is the result of a proton inside the nucleus changing into a neutron, whereas normal β-decay (negative electron emission) involves the change of a neutron into a proton. These processes are written as follows, using the notation n for neutron, p for proton, ν for neutrino and $\bar{\nu}$ for antineutrino:

$$n \rightarrow p + \beta^- + \bar{\nu}$$
$$p \rightarrow n + \beta^+ + \nu$$

They normally occur only in the nucleus, as for example in the case of ^{64}Cu, which, unusually, can undergo both types of β decay:

$$^{64}\text{Cu} \rightarrow {}^{64}\text{Zn} + \beta^- + \bar{\nu},$$
$$^{64}\text{Cu} \rightarrow {}^{64}\text{Ni} + \beta^+ + \nu.$$

However, the decay of the free neutron is energetically possible, and it does decay, with a half life of 10 minutes. The neutron is heavier than the combined masses of its decay products. The missing mass reappears as energy of the emitted particles, in accordance with Einstein's special theory of relativity, published in 1905; $E = mc^2$, as every schoolchild knows!

Predicted theoretically by Paul Dirac in 1928 and first observed in

1932 by Carl Anderson, in cloud-chamber studies of cosmic rays, the positron is a form of antimatter. The electron and the positron form a particle/antiparticle pair, as do Pauli's neutrino and the antineutrino; all are categorized by particle physicists as 'leptons', the above equations illustrating lepton conservation.

A positron readily combines in a spectacular manner with an electron in a process known as 'annihilation' in which both particles disappear to be replaced instantly by two high-energy photons of equal energy which are emitted in opposite directions: $E = mc^2$ with a vengeance! This effect, as we shall see, has been exploited in nuclear medicine. The high energy photons from positron annihilation are very similar to γ rays from radioactive decay.

In 1934, based upon Pauli's neutrino hypothesis, to which reference has already been made, Enrico Fermi produced a theory of β decay which reproduced exactly the observed distribution of β particle energies. We now know that β decay is a consequence of the so-called weak force. α decay is conceptually simpler. Nucleons tend to cluster together inside the nucleus as α particles. In some very heavy nuclei the nuclear force is not strong enough to hold everything together and the result is emission of an α particle, leaving a more stable nucleus.

This is about as far as I wish to go in discussing the development of our understanding of radioactivity. We have reached the point at which a clear picture has emerged of the nucleus as a conglomeration of protons and neutrons. From this starting point, and taking on board the equivalence of mass and energy, nuclear physicists have developed a greater understanding of how these particles are bound together in stable nuclei and why some nuclei are unstable and decay by α- or β-ray emission.

In 1938 Otto Hahn showed that barium could be produced by bombarding uranium with slow neutrons, the process called neutron-induced fission, and the scene was set for the development of nuclear weapons and civil nuclear power. At the same time, the growing availability of artificial radionuclides led to a massive expansion in the applications of radioactivity. I shall now consider some of these applications, returning first to radium.

Medical applications of radium

X rays were being used for therapeutic purposes within a year of their discovery, but medical applications of radioactivity took longer to establish. In 1901 Henri Becquerel made his last major contribution to the subject of radioactivity when he reported a burn to his skin after having carried a radium source in his shirt pocket for some six hours. Rumour has it that he checked this discovery by transferring the source to another pocket for a few hours and made absolutely sure by persuading his friend Pierre Curie to repeat the experiment!

The effect noted by Becquerel was recognized to be identical to X-ray dermatitis and led quickly to the use of radium for treating lupus and other skin conditions. These treatments were successful and sealed tubes and needles containing radium, or radon, became widely used in the treatment of surface, interstitial, and intracavitary tumours. In many cases the radium source was in close contact with the tumour, a form of treatment called brachytherapy, as distinct from teletherapy which uses beams of radiation from a source at a distance.

For a while, radium was regarded as a 'cure-all', as shown by advertisements dating from around 1913, such as that reproduced in Fig. 7. The apparatus depicted dispensed radioactive drinking water, recommended as a treatment for gout, arthritis, and a host of other afflictions: such treatments, involving internal administration of radium, were abandoned when they were shown to do more harm than good, due to the deposition of radium in bones and the subsequent effects of α radiation.

By the 1920s the hazards associated with radium were better understood, advice was available on safe handling, and applications became restricted largely to the treatment of malignant diseases. During the 1920s teletherapy units, each employing several grams of radium in a thick lead container with a collimating aperture, were developed for γ-ray therapy. In the 1930s elaborate procedures were developed for determining the optimum therapeutic dose distribution from arrays of radium tubes and needles. Among the best known of these was the so-called Manchester system, which remained in use until at least the late 1960s.

In 1929 the National Radium Trust (NRT) was established to oversee the use of radium in UK hospitals. The trust received £250 000 from a fund established by *The Times* newspaper to give thanks for the recovery of King George V from a long and serious illness, and this was used to purchase some 50g of radium.

In due course the holding of substantial amounts of radium was organized around certain hospitals acting as regional centres, with the National Physical Laboratory (NPL) responsible for the examination and testing of all radium containers. In the year 1930 NPL tested no less than 3000 individual containers, mostly for the NRT. It was then the practice to transport up to 50 mg of radium at a time by the ordinary post, using a 2.5 cm thick lead pot inside a wooden box. Larger quantities, up to 5 g, were packed in a lead pot of maximum wall thickness 5 cm and taken

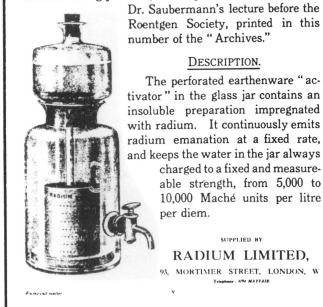

RADIUM THERAPY

The only scientific apparatus for the preparation of radio-active water in the hospital or in the patient's own home.

This apparatus gives a <u>high</u> and <u>measured</u> dosage of radio-active drinking water for the treatment of gout, rheumatism, arthritis, neuralgia, sciatica, tabes dorsalis, catarrh of the antrum and frontal sinus, arterio-sclerosis, diabetes and glycosuria, and nephritis, as described in Dr. Saubermann's lecture before the Roentgen Society, printed in this number of the " Archives."

DESCRIPTION.

The perforated earthenware " activator " in the glass jar contains an insoluble preparation impregnated with radium. It continuously emits radium emanation at a fixed rate, and keeps the water in the jar always charged to a fixed and measureable strength, from 5,000 to 10,000 Maché units per litre per diem.

SUPPLIED BY

RADIUM LIMITED,

93, MORTIMER STREET, LONDON, W

Telephone: 4796 MAYFAIR

Fig. 7 Apparatus for producing radon-bearing drinking water, recommended as a curve for numerous disorders; an advertisement of circa 1913. (Reproduced from R.F. Mouild, 1980, by kind permission of Reid Business Publishing Ltd.)

by rail. By the end of 1936, 130 g of radium had been tested at NPL, estimated to be 20 per cent of the world's supply.

With the setting up of the National Health Service in 1948 the NRT and its associated Commission were wound up and routine testing was transferred to the Radiochemical Centre at Amersham. Thereafter, with the increasing availability of artificial radionuclides, the use of radium declined and it is now no longer used in UK hospitals. γ-ray teletherapy units are nowadays sourced with ^{60}Co or ^{137}Cs, while ^{192}Ir is now the most widely used radionuclide in brachytherapy.

Radium standards

Having introduced the role of the NPL in testing radium sources, this is perhaps the appropriate point at which to say something about early radioactivity standards and how NPL came to be involved. I have already spoken about the electrical method for measuring radioactivity. When this method came to be applied to measurements extending over long periods of time, it was recognized that a method of checking the stability of the electrometer was needed, and by 1906 the preparation of standard sources of uranium oxide had been described.

In that same year, 1906, the Röntgen Society, meeting in London and recognizing that a radium sample is a far more stable radiation source than an X-ray tube, decided upon the need to establish a radium γ-ray standard to serve as a basis for quantifying doses administered in X-ray therapy, and for the standard to be kept at the NPL. A committee was formed and determined that the standard should be based upon the γ-ray emission from 1 g of pure radium bromide filtered by 5 mm of lead. This UK initiative was, however, overtaken by international developments.

In 1910 the International Radium Standards Commission was created, with Rutherford as its president, for the express purpose of establishing an international radium standard. This standard was duly prepared by Marie Curie: it comprised, as of August 1911, 21.99 mg of pure radium chloride sealed into a thin-walled glass tube. This standard was compared satisfactorily with three similar standards prepared in Vienna by Otto Honigschmid in the course of redetermining the atomic weight of radium. The Commission arranged for further standards to be prepared and calibrated against the Paris and Vienna standards; one of these secondary standards, no. 3, became the British National Radium Standard, and was finally deposited at NPL in June 1913, after having been inspected by Marie Curie in Paris. It was stated to be equivalent, in 1912, to 16.08 mg of radium element.

In 1934 the original British National Radium Standard was replaced by a new standard of similar form (see Fig. 8), one of a set prepared by Otto Honigschmid for a fresh determination of the atomic weight of radium; a comparison of the old and new standards showed that they agreed to within 0.3 per cent. Besides the UK, a number of other countries acquired the new Honigschmid samples and adopted them as national standards; the Paris and Vienna standards of 1912 were also replaced. During the 1950s there were several intercomparisons of these standards and in 1960 the International Committee of Weights and Measures formally adopted the 1934 Honigschmid samples as the international

Fig. 8 British National Radium Standard, 1934, with an ionization chamber previously employed at NPL for the measurement of radium sources.

standard for the future measurement of radium: at the same time the International Bureau of Weights and Measures, BIPM, at Sèvres assumed responsibility for the Paris standard. It is interesting to note that only two international measurement standards now exist which are based upon physical artefacts; one is radium, the other is the international standard kilogram, also held in Sèvres.

The replacement of the 1912 standards had been motivated by concern over the integrity of the fragile glass containment. A similar concern led NPL to establish, in 1981, a set of three secondary standards comprising anhydrous radium bromide encapsulated in robust, platinum–iridium tubes, in order to maintain traceability of radium measurements to the international standard for the foreseeable future. One of these secondary standards has been calibrated at the National Institute of Science and Technology (NIST) in the USA (formerly the National Bureau of Standards, NBS) and at the D.I. Mendeleev Institute in the former Soviet Union. In these latest intercomparisons three different measurement techniques were used; at NPL a new technique was used, based on measuring γ rays from ^{214}Po with a modern γ-ray spectrometer, while the other laboratories employed an ionization chamber and/or a sensitive calorimeter, both well-tried and tested methods.

With the decline in the medical uses of radium the maintenance of radium standards is now justified largely by the importance of radon as the largest single source of the background radiation to which we are all exposed (see below).

Quantities and units

The advertisement shown in Fig. 8 refers to the strength of the radio-activity in Maché units, and this serves as a reminder of the need to give further consideration to quantities and units in radioactivity. With the establishment of radium mass standards the effects of medical exposure to radium could be quantified in terms of the amount of radium and the duration of the exposure; a milligram-hour unit was defined in 1912. The curie unit was defined at around the same time, as an aid to quantifying the effects of radon gas, for which a mass standard was not feasible: the curie was defined as 'the quantity of radon in radioactive equilibrium with 1 g of radium element'. The curie corresponded to 6.5 μg of radon and represented a decay rate of approximately 3.7×10^{10} disintegrations per second, d.p.s., of radium or radon.

To provide a measure for other radionuclides, the rutherford unit was proposed in 1946 as 'the amount of any radioactive isotope which decays at the rate of 10^6 d.p.s.. However, the rutherford never became widely used, and in 1953 the curie was recommended as the unit of activity for any radionuclide and was redefined to be 3.7×10^{10} d.p.s. exactly; the use of the rutherford was discontinued. Finally, in 1975, the becquerel (Bq), equal to one disintegration per second, was defined as the unit of activity within SI, the International System of Units. It was recommended that the curie should be phased out, and in the UK 'The Units of Measurement Regulations 1980' specified that the curie, along with the rad, the rem, and the röntgen, should cease to be used as from 1 January 1986.

As recently as 1980, the International Committee for Radiation Units and Measurements (ICRU) finally reached agreement upon a formal definition of activity, as follows:

> The activity of an amount of a radioactive nuclide in a particu-
> lar energy state at a given time is the quotient of dN by dT,
> where dN is the expectation value of the number of sponta-
> neous nuclear transitions from that energy state in the time
> interval dT.

This definition recognizes that radioactive decay is a statistical process and states in effect that the activity of a sample is the average rate of disintegration within it; for example, in a ^{60}Co sample of activity 1 MBq, an average of 10^6 Co atoms decay per second.

Dectectors

At this juncture it may be helpful to consider the development of devices for detecting and measuring radioactivity. Most of the detectors used today have their origins in one or other of two devices which we have already met, the gas ionization chamber and the scintillating ZnS screen, although the ionization chamber is still widely used in its original form, in which the current measured corresponds exactly to the rate at which charge is released by the incident radiation.

A very important development of the ionization chamber was the gas-flow proportional counter, in which a strong electric field facilitates additional ionization by collision, so that each incident α or β particle gives rise to a substantial electrical signal or pulse whose amplitude is proportional to the energy deposited in the gas. The Geiger counter, invented in 1908 and improved by Geiger and Müller in 1928, is similar to the proportional counter but is operated under conditions such that the output pulse is of fixed amplitude regardless of the incident radiation.

As regards scintillation detectors, in 1944 a ZnS screen was coupled to a photomultiplier tube for the first time, so that electrical counting could replace the human observer. Over the next few years a range of organic scintillators was developed, along with the ubiquitous thallium-activated sodium iodide detector for γ rays. More recently, liquid organic scintillators have been applied with great success to the measurement of low-energy β emitters such as ^{3}H (tritium) and ^{14}C.

The modern semiconductor detectors, dating from the 1950s and now widely used for X- and γ-ray spectrometry, are ionization devices in which the gas has been replaced by a solid semiconducting material, most commonly high-purity germanium. These detectors are highly efficient and have good energy resolution: they are, however, relatively expensive and must be operated at liquid nitrogen temperatures. The characteristic line spectra recorded from γ emitters by such instruments act as unique fingerprints, facilitating the identification and measurement of individual radionuclides (see Fig. 9).

More about standards

The rapid growth in the use of artificially produced radionuclides after World War II demanded the development of appropriate standards of radioactivity. To this end, NPL quickly established informal links with interested organizations in the UK, including the Atomic Energy Research Establishment at Harwell, the Medical Research Council Unit

Fig. 9 γ-ray spectrometry being used at NPL for the measurement of environmental level radionuclide standards.

at Hammersmith Hospital, and the Royal Cancer Hospital. Comparisons of measurements were organized between these bodies and with the National Bureau of Standards in the USA and the National Research Council in Canada. In 1953 NPL was formally recognized as the body responsible for all UK radioactivity standards.

Over the years which followed there was a steady growth in standards work, which acquired an increasingly international flavour. In 1958 the International Committee of Weights and Measures established a Consultative Committee for Standards of Measurement of Ionizing Radiations (CCEMRI, from the French). In 1964 new laboratories for this work were dedicated at the BIPM, and in 1969 three sections were established within CCEMRI, one of which, Section II, is concerned wholly with radioactivity measurements and in particular with ensuring international agreement on primary standards.

Because of the transient nature of radioactivity, primary standards, except for radium, are based upon a measurement method rather than a physical artefact, and the becquerel must be separately realized for each radionuclide. Primary standards of radioactivity are now determined largely by the method of 'coincidence counting', which takes advantage of the fact that most radionuclides of interest decay by the simultaneous emission of β and γ rays. For example, Fig. 10(a) shows the simplest decay scheme (as applying, for example, to ^{203}Hg). As usual, the γ rays arise from the very rapid de-excitation of the 'daughter' nucleus produced by β decay. In general there may be several β branches and numerous γ rays.

In coincidence counting (Fig. 10(b)), the source to be measured is exposed simultaneously to a β detector and a γ detector. Neither detector need have a detection efficiency of 100 per cent, but the coincidence unit, functioning as an AND gate, counts the number of times that the two detectors register an event simultaneously. From the observed

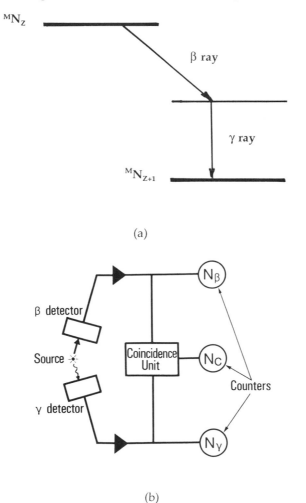

(a)

(b)

Fig. 10 Determination of activity by coincidence couting: (a) simple radionuclide decay scheme, (b) schematic diagram of measuring system.

counting rates, N_β, N_γ, and N_c, we can calculate the activity, N_o, of the source, also the detector efficiencies. In general the aim is to standardize radionuclides in aqueous solution: in this case drops of the solutions are weighed out onto very thin, conducting foils and evaporated to dryness to provide sources for measurement.

Figure 11 shows equipment used at NPL for coincidence counting; it features an automatic sample changer and is fully computerized. Heavy lead shielding protects the detectors from extraneous radiations.

Radioactivity measurements are dominated by statistics and the need to count for a long time in order to get a precise result. In practice there are other difficulties; decay schemes are complicated, equipment is imperfect, half lives and other nuclear data may not be well known, radionuclide impurities may sometimes be present, and background radiation is always present. Each new radionuclide brings new measurement problems, and validation of new and improved measurement techniques through international intercomparisons remains a crucial requirement; such intercomparisons, mostly under the auspices of the BIPM, are held about every two years.

Once they have been established, primary standards of radioactivity are maintained and disseminated though secondary standard sources or

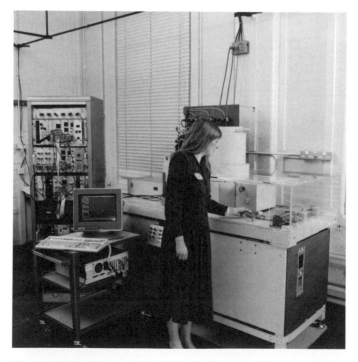

Fig. 11 Equipment used at NPL for β-γ coincidence counting.

instruments. Radioactive standard sources are available from NPL and elsewhere, while the NPL secondary standard radionuclide calibrator provides simple and accurate measurements of a range of important radionuclides at MBq levels of activity, traceable to primary standards.

Nuclear medicine

Having established that a robust measurement system exists for radio-activity, providing the means to maximize the benefits and minimize the risks associated with its use, we may now return with more confidence to a further consideration of some of these applications.

There is no doubt that one of the most benign consequences of Henri Becquerel's discovery has been the development, very largely since the last war, of nuclear medicine, defined as 'the diagnostic and therapeutic applications of unsealed radionuclides administered to the patient in vivo'.

As regards therapeutic applications of nuclear medicine, one thinks naturally of the treatment of malignant tumours, but benign diseases are also treated, for example hyperthyroidism, synovitis, and polyeythaemia.

One of the earliest examples of diagnostic nuclear medicine was the use of ^{131}I to study thyroid function. Since then, many different radio-nuclides have been applied, to a wide range of investigations. Figure 12, dating from the 1960s, indicates both the versatility of the method and the range of radionuclides which have been used.

These days, however, most work is done with a handful of popular radionuclides, of which ^{99m}Tc, first introduced in 1958, is by far the most widely used. There are now some half million administrations of radionuclides to patients in UK hospitals each year, of which more than 97 per cent are for diagnosis.

For diagnostic nuclear medicine the ideal radionuclide would emit a single γ ray with just enough energy to escape from the body and be detected, it would have a high chemical affinity for the organ or process under study and a short physical or biological half life to minimize the radiation dose to the patient.

In the early days the movement of isotopes within the body was tracked with simple detectors on a point-by-point basis, but by the 1960s Hal Anger had perfected the γ camera. This is a very large sodium iodide scintillation crystal, fitted with a number of photomultiplier tubes to provide spatial resolution, used in conjunction with a focusing collimator. The collimator, of lead, is provided with a conical array of holes, so arranged that the detector can receive only those γ rays emerging from a well defined, small volume within the subject.

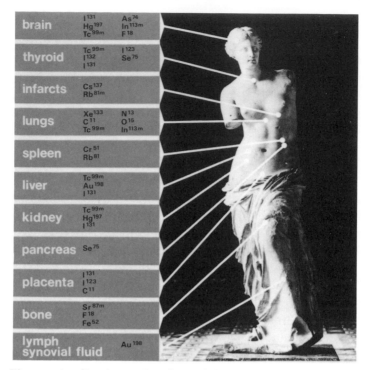

Fig. 12 Applications of radionuclides in nuclear medicine. (From a photograph kindly provided by the Louvre, Paris.)

No account of nuclear medicine would be complete without at least a mention of tomography, the method of imaging which provides pictures in the form of slices through the body. By the 1960s X-ray tomography was well advanced and similar results were obtained in nuclear medicine by rotating a γ camera around the patient, the so-called method of single photon emission tomography or SPET. The early tomographic images were blurred. Godfrey Hounsfield and Alan Cormack showed how computers could be used to eliminate the blurring: they received the Nobel Prize for physics in 1979 for thus inventing computed tomography or CT.

I have already referred to the usefulness of positron-emitting radionuclides in nuclear medicine; Fig. 13 shows one way of using this effect. An arrangement for positron imaging is shown on the left; a non-tomographic image is obtained by recording events which are registered simultaneously by a γ camera, used without a collimator, and a small scintillation counter.

Comparison with X-ray imaging, on the right of Fig. 13, shows that the geometries are identical, and in each case tomographic images result from rotating the system around the patient. Positron emission tomogra-

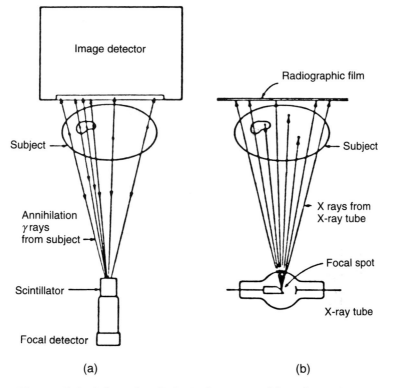

Fig. 13 Principles of radiological imaging: (a) with positrons and (b) with X rays. (Reproduced from S. Webb, 1990, by kind permission of IOP Publishing Ltd.)

phy, or PET, has now become a particularly powerful imaging technique: in the UK, the Medical Research Council's Clinical Sciences Centre at the Hammersmith Hospital has played a major part in the development and application of PET. Modern PET scanners use many small detectors rather than a γ camera. The most popular radionuclides for PET are short-lived isotopes of carbon, oxygen, and fluorine. Plate 3 illustrates the amount of detail which a PET scan can reveal.

These days, X-ray and radionuclide imaging compete with the newer methods of ultrasound and magnetic resonance imaging; however, nuclear medicine techniques have individual strengths sufficient to ensure that they will continue in use for many years to come.

Other applications of radioactivity

In addition to medicine, radioactivity finds application in numerous other areas. In biochemistry, radioactive isotopes of hydrogen, carbon, sulfur, etc., are incorporated into labelled compounds for use as tracers. Thus, for example, labelling of a new insecticide allows its uptake and distribution to be followed, not only within the target species but in the environment at large. Other life sciences applications include DNA sequencing and drug profiling.

In geophysics, early measurements of the relative proportions of uranium isotopes and their decay products in rocks confirmed that the age of the Earth is about 4600 million years, conflicting somewhat with the estimate in 1664 by James Ussher, Archbishop of Armagh, that the creation occurred in the year 4004 BC! Similar measurements of the decay of the naturally occurring radionuclide ^{40}K to stable ^{40}A have enabled a wider range of rocks to be dated with improved precision, and the use of radiocarbon dating for the dating of archaeological artefacts such as the Turin Shroud is well known.

In modern industry the scattering and absorption of radiation from radioactive sources are widely used for process monitoring and control; examples include liquid level measurement and thickness gauging. In radiography, radionuclide sources are used to investigate the integrity of structures such as aircraft engines, bridges, and pressure vessels, while activation techniques are used to study wear and abrasion in mechanical systems. High-activity sources find application in radiation processing and sterilization, e.g. of single-use medical products. The manufacture and supply of radionuclide-based products are themselves activities of substantial economic importance.

Radiation protection

All of these applications of radioactivity are now permitted only in strict compliance with internationally established principles of radiation protection, based upon past experience and new knowledge. In the UK, the National Radiological Protection Board, NRPB, has a key role in advising government upon matters relating to radiological protection.

Proof of compliance with regulations requires sound measurements, traceable to national standards. Radionuclide metrology is now a worldwide activity and the number of radionuclide metrologists today probably exceeds the total number of physicists of all complexions at the time of Becquerel's discovery!

In 1959 and 1966 the International Atomic Energy Agency, IAEA, had organized international symposia on radioactivity measurements in Vienna, and in 1972 there was a particularly successful summer school at Herzeg Novi in the former Yugoslavia: this led to the establishment of the International Committee for Radionuclide Metrology, ICRM, an informal organization which aims to foster collaboration in all aspects of applied radioactivity measurements. Week-long meetings are held every two years and attract increasing numbers of delegates.

The most serious consequences of exposure to radioactivity are associated with the inhalation or ingestion of α-emitting radionuclides. Although α particles have a very short range in tissue, they produce very dense ionization and cause serious damage to cells; they are some 20 times more effective in this respect than β particles or γ rays. Examples of the effects of exposure to α rays can be found among uranium miners, early researchers into radioactivity, and the unfortunate dial painters who licked their paint brushes in the course of applying radium-based luminous paint to watches and other instruments.

Today, the only significant exposure to α radiation for most people arises from naturally occurring radon gas. This point is emphasized by the pie-chart produced by the NRPB (Fig. 14). This shows the various contributions to the average yearly dose of 2.6 mSv (millisieverts) received by members of the public in the UK.

Some 87 per cent of our average annual dose comes from natural sources, quite independent of Becquerel's discovery: the 50 per cent con-

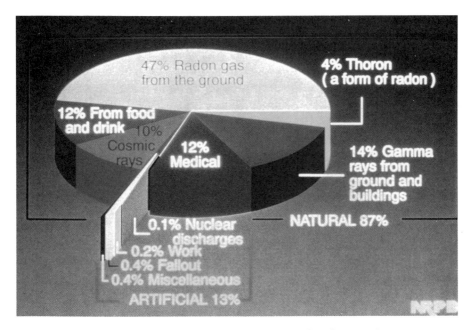

Fig. 14 Components of the average annual radiation dose to members of the public in the UK. (Reproduced by kind permission of the National Radiological Protection Board.)

tribution from radon is estimated to be responsible for 6 per cent of all lung cancers. Radon levels in some places, notably in Cornwall and parts of Scotland, can be much higher than average and Government advice and assistance is available in such cases, where the risk is deemed to be particularly significant.

It may be noted that the contribution from the nuclear industry is very small. The excellent record of the UK nuclear industry in limiting the radiological consequences of its activities is due in no small measure to comprehensive environmental monitoring programmes carried out by the operators and the regulatory bodies. NPL contributes to this work through the provision of environmental radioactivity standards and measurement intercomparisons.

The future

As we move now towards the Becquerel bicentenary it is clear that his discovery will continue to play an important part in all our lives. The applications of radioactivity in medicine, industry, and research will continue to increase and so too, I believe, will our dependence upon nuclear power.

While research will continue upon energy generation from the fusion process in light nuclei, practical power stations based upon this process are unlikely to be realized within about 50 years from now. Europe now generates almost one third of its electricity from nuclear fission; for the UK the figure is 25 per cent. It seems inconceivable that this dependence will diminish.

The excellent safety record of the UK nuclear industry has already been mentioned: the main cause of continuing public concern is the disposal of spent nuclear fuel and other nuclear waste. This waste is categorized in terms of its activity as (a) low level, lightly contaminated material from a variety of sources and comprising 90 per cent of all UK nuclear waste, (b) intermediate level, more highly contaminated material, from the nuclear industry, and (c) high level, comprising fission product waste from the nuclear fuel reprocessing plant.

Nuclear waste disposal was subject to government review in 1994 and a white paper was published in July 1995. Its conclusions were that low-level waste should continue to be disposed of via shallow burial sites such as Drigg in Cumbria and in landfill sites, and that high-level waste should be subject to vitrification and storage for 50 years followed by deep disposal. For the most difficult category, intermediate-level waste, early deep disposal is the chosen option, consistent with the policy of sustainable development, but only after further research and a full public enquiry. Some 10 000 tonnes of intermediate-level waste are produced in the UK each year; this may seem a great deal, but should be compared with the more than 5 million tonnes of toxic chemical waste produced annually. The nuclear industry is optimistic that a repository for deep disposal will be built in due course at Sellafield, in Cumbria. A major task facing the industry in achieving this objective will be to win over the hearts and minds of the public. This will have been helped by the creation in April 1996 of the Environment Agency, which brings responsibility for matters radiological under the same umbrella as other factors affecting the environment.

What of the new science of nuclear physics which emerged as a consequence of Becquerel's discovery? In many ways the 1930s and 1940s were the golden age of nuclear physics. In recent times the UK nuclear physics community has continued to make major contributions to the subject, despite reduced funding and wholesale closure of national facilities. However, many fundamental questions about the nucleus remain unanswered.

One current hot topic which may answer some of these questions is the search for nuclei at the limits of stability. As recently as February 1996, an international team working at the GSI laboratory in Darmstadt, Germany, discovered its sixth element, producing the heaviest nuclide to date, element no. 112, atomic mass 277. This was achieved using a beam of ^{70}Zn ions incident upon a ^{208}Pb target. ^{70}Zn is a stable isotope, and further substantial progress in this field will require beams of radioactive ions. There is guarded optimism amongst UK nuclear physicists that a European Radioactive Beam Facility may be established at the existing ISIS facility of the Rutherford Appleton Laboratory at Harwell.

Conclusion

In this brief review of Henri Becquerel's discovery and its consequences, an attempt has been made to bring out the importance of measurement. It has, therefore, seemed appropriate to seek a measure by which one might justifiably quantify the impact of Becquerel's discovery upon science and society at large. Several references have already been made to Nobel Prizes, and further investigation reveals that, in the first 75 years since the discovery of radioactivity, more than 40 scientists won Nobel Prizes in physics or chemistry for work either directly related to radioactivity or dependent upon its discovery. This remarkable fact should stand as a lasting tribute to the importance of the findings reported by Henri Becquerel to the French Academy of Sciences on 2 March 1896.

Acknowledgements

I am indebted to NPL Management Ltd for facilitating the preparation of this discourse and to AEA Technology with UK NIREX Ltd, Amersham International plc with the Royal Society of Chemistry, the National Radiological Protection Board, the MRC Clinical Sciences Centre, Hammersmith Hospital, and former colleagues of the Centre for Ionizing Radiation and Acoustics, NPL, for their contributions to the exhibition that accompanied the lecture upon which this chapter is based.

Further reading

E. Rutherford, *Radioactive Substances and their Radiations*, Cambridge University Press, 1913.

F.W. Aston, *Isotopes*, Edward Arnold and Co., London, 1922.

Radium – Production – General properties – Therapeutic Applications – Apparatus, Union Minière du Haut Katanga, Brussels, 1929.

E. Rutherford, J. Chadwick, and C.D. Ellis, *Radiations from Radioactive Substances*, Cambridge University Press, 1930.

O. Glasser (ed.), *The Science of Radiology*, Balliere, Tindall, and Cox, London, 1933.

A. Romer, *The Restless Atom*, Dover, New York, 1960.

A. Romer, *The Discovery of Radioactivity and Transmutation*, Dover, New York, 1964.

A. Romer, *Radiochemistry and the Discovery of Isotopes*, Dover, New York, 1970.

C.H. Page and P. Vigoureux (eds), *The International Bureau of Weights and Measures 1875–1975*, National Bureau of Standards Special Publication 420, US Government Printing Office, Washington DC, 1975.

E.E. Smith, *Radiation Science at the National Physical Laboratory 1912–1955*, HMSO, London, 1975.

R.F. Mould, *A History of X rays and Radium*, IPC Business Press Ltd, London, 1980.

F. Close, *The Cosmic Onion—Quarks and the Nature of the Universe*, Heinemann Educational Books Ltd, London, 1988.

S. Webb, *From the Watching of Shadows—The Origins of Radiological Tomography*, Adam Hilger, Bristol and New York, 1990.

A.T. Elliott, F.M. Elliott, and R.A. Shields, *Nuclear Med. Comm.*, 1993, **14**, 360.

Review of Radioactive Waste Management—Final Conclusions, Cm2919, HMSO, London, July 1995.

S. Quinn, *Marie Curie—A Life*, William Heinemann Ltd, London, 1995.

J.S. Laughlin (ed.), Origins of the Science of Radiation Physics and of the Field of Radiology, *Medical Phys.*, 1995, **22**, no. 11, part 2.

P.I.P. Kalmus, Is matter empty? *Science and Public Affairs*, Royal Society and British Association, autumn 1995.

Review of Nuclear Physics—Report of the Review Panel Chaired by Professor A C Shotter FRSE, Engineering and Physical Sciences Research Council, Swindon, March 1996.

PETER CHRISTMAS

Born 1935, he received his D.Phil. from Oxford University in 1962. After three years at the National Physical Laboratory as a Senior Research Fellow, and two years at the National Research Council, Ottawa, as a Postdoctoral Fellow, he returned to NPL in 1967 to pursue a career in radionuclide metrology, remaining at the Laboratory until his retirement in December 1995 as Director, Centre for Ionizing Radiation and Acoustics. He is past President of the International Committee for Radionuclide Metrology.

Alcohol and the development of human civilization

BERT VALLEE

Alcohol, a Substitute for Water as the Major Human Beverage

Since the beginning of Western civilization beer and wine *not* water have been the major sources of human liquid intake. The fermented products of grapes designated 'wines' succeeded the products of fermented cereals, collectively referred to as 'beer', a few thousand years later. Most likely both beverages resembled those presently known by these names only remotely. Their alcohol content was characteristically variable and low. Beer and wine in whatever form completely dominated the consumption of human beverages for the first 16 000 years of known Western history. Not until 1100 AD, i.e., the beginning of the current millennium, did the results of distillation, e.g., brandy, liqueurs, gin, rum, and/or whiskey make their debut.

Since antiquity and until 1900 AD Western society considered water impotable. Egyptian, Babylonian, Hebrew, Assyrian, Greek, Roman, and subsequent Western societies all rejected it as a beverage. It was known to cause acute and/or chronic but deadly illnesses, was poisonous and to be avoided, particularly when stagnant[1]. Both the Old and New Testaments hardly refer to water as a beverage as is true for Egyptian, Greek, and Roman sources as well, exempting only water from mountain springs or cisterns. The want of liquids safe for human consumption prevented and curtailed long range voyages.

Microbiology which ultimately identified bacteria as the source of the ill effects of water pollution remained unknown, of course, until the nineteenth century AD. It is now clear that pathogenic bacteria cannot survive the pH and ethanol content of wine regardless of how it is made. The accompanying sterility is what made wine making and drinking so important in Southern Europe. Wine was a safe drink when water constituted a potential, serious health hazard[2]. Neither boiling of water with the express purpose of destroying bacteria and other parasites nor other purification methods were employed in Western civilizations.

The experience in the Far East, especially China differed vastly. Boiling of water to brew tea was already common practice and very popular in China during the Han dynasty (~200 BC) (Lu Yum, ~800), and there were explicit references to the cultivation, harvesting and growth of tea and its preparation with boiled water but no comment on possible effects of water on health. In ancient China, tea was first used by the royal and noble families ~1000 BC but it became popular 800 years later as a thirst quencher and a safe supply of liquids for the masses. In Western society, for nearly 10 000 years beer and wine, *not* water, were the major daily thirst quenchers consumed by all ages, and boiling of water to brew coffee was delayed until 1650.

Seafaring nations and explorers knew that water becomes putrid after very short periods of storage. Beer or wine seem to have been the only alternatives available then. Thousands of years later during their epic battle the British and Spanish fleets carried beer and wine respectively as the sole beverages; their logs do not even mention water. Those of the Mayflower indicate that she had run out of beer for the passengers when she landed in Plymouth Harbor in 1621. Desperate, the passengers were astonished to find the local sources of water acceptable, but remained intent on obtaining beer instead as the potable liquid that they trusted. Almost immediately after arrival, the settlers focused attention on luring brewmasters to the colonies to ensure local production of beer. Brewing of beer became a household chore still practised by George Washington and Thomas Jefferson in the eighteenth century.

Origins of beer and wine

Cereal fermentation resulted in 'beer' which was probably first brewed between the sixth and eighth millenia BC coincident with the production of grain crops in Africa (Fig. 1)[3]. The cultivation of grapevine added wine to the liquid resources of neolithic man. Wild grapevine was indigenous to nearly all present great wine-producing countries of Europe and the Near East. Its cultivation has been attributed to an early neolithic race in Transcaucasia (probably Armenia) around 6000 BC. Wine and beer supplied and replaced body fluids and quenched thirst but milk was considered 'barbaric'. The waters of the Indus, Nile, Tigris, and Euphrates rivers were thought to be potable, in contrast to all other

Malting

Brewing

Fig. 1 Beer-making in Egypt, 2500 BC.

sources in that region, exempting water from mountain springs and rain water collected in cisterns. From antiquity to modern times, virtually all social systems held beer and wine in high esteem, much as from time to time concerns were voiced about the consequences of their excessive consumption but, most likely, their alcohol content was so low that this caused few, if any, ill effects.

In retrospect, most likey the relatively large amounts of acetic and other acids as well as additional products of fermentation, contributed to the effects of their mixture with alcohol to lower the content of pathogens in water. Moreover, for nearly 5000 years of recorded history the 'sour' taste of beer was commented on perpetually.

Wine was a luxury not accessible to the masses. Most likely it more closely resembled either vinegar or syrup rather than wine that was prepared and reserved for medicinal and sacred purposes of the ruling classes (kings, courtiers, nobility, and the priesthood) (Fig. 2).

By and large, those beers and wines were not very intoxicating; their thirst and hunger quenching characteristics far overshadowed their effects on mood and were insignificant relative to their benefits. Excessive consumption or tolerance took back seats to extensive debates on suitable *dilutions* of beer and wine with water in order to adjust either taste or cost which were debated and discussed constantly for more than 2000 years. There seems to have been nearly universal agreement that the proper proportion of the mixture should be one part of beer or wine diluted with two parts of water.

Fig. 2 Wine-making: pressing grapes in a bag (from Mereruka's tomb, 6th Dynasty Egypt, 2423 BC).

The oldest physically preserved 'pharmacopeia', a Sumerian Clay Tablet in cuneiform script dating from ~2100 BC describes the production of beer and has been cited as the first record of the therapeutic use of alcohol[4]. While the historical accuracy of a Persian legend is doubtful, it ascribes the first medicinal use of alcohol to the emperor Djernshid (~800 BC), the founder of Persepolis. He decreed that grapes stored in jars for the winter and believed to have become spoiled and poisonous be given to criminals for capital punishment. The unexpected consequences of the consumption of the supposed poison persuaded the emperor to use it instead as an encouragement for troops in battle!

A component of the diet and source of calories

In their zeal to define alcohol as a scourge of human existence, historians may have underemphasized or overlooked that it was also a major source of essential calories. The fertile river deltas of Egypt and Mesopotamia produced huge crops of wheat and barley which provided beer yielding 4 and 7 kcal/gm when consumed as carbohydrates or alcohol, respectively. According to Egyptian records, while the common man also had access to some animal protein, the amount was clearly insufficient for conditions which called for high energy output, e.g., hard manual labour, marching or rowing. Records recovered from most ancient civilizations document the life style of the elite, but not that of the peasants or labourers whose existence was marginal. In addition to alcohol and carbohydrates beer also provided essential micro nutrients such as vitamins and minerals.

The 'side' effects of beer and wine accounted for the 'normal' state of mind of both the young and old. Beer and wine induced a sense of well-being counteracting both the prevalent state of fatigue and boredom and alleviating wide-spread pains of all types for which there were no known remedies. Together with food, alcohol encouraged socialization while providing one of the few sources of enjoyment and release of tensions. From early on both beer and wine were important aspects of religious and civic festivals, feasts and community affairs, the basis for the observance of Dionysian and Bacchanalian rites, and later on religious holidays and festivities of the Church, i.e., including Carnivals, Circuses, Christmas, Easter, Pentecost, and Saints Days.

Quite the same, the results of excessive alcohol consumption were well known and described and subject of debate on the part of philosophers and religious leaders. Early on, Plato, Socrates, and Xenophon expressed opinions on the subject. Xenophon (431–351 BC) gave a classic description of the clinical effects of 'wine', which is virtually identical with that known presently: a pleasant thirst quencher inducing joy, cheer, and good humour, counteracting grief, promoting and cultivating friendships, fostering gaiety, a central feature of hospitality; he recognized its adverse effects on diligence and memory, potentially causing anger, and violence.

In spite of the remarkable political and religious upheavals during nearly 3000 years in the Western world, the consumption of, and reaction to, beer and wine during the rise and fall of the Roman Empire, the migrations of the Germanic tribes and the succeeding Dark Ages with their accompanying religious and political realignments remained much the same. Contrary to some modern perceptions, the Catholic Church did not take a militant stand against alcohol. To the contrary, it was both its primary producer and steady consumer. Early versions of the New Testament hardly refer to alcohol, its use or abuse. Christ may or may not have known about the adverse effects of wine since its availability was largely limited to the upper classes. He clearly approved of its consumption and established it as the symbol for penitence and the remission of sins through the Eucharist and was renowned for the miracle of converting water into wine. Overall, the apostles were more concerned with measures to control the use and excessive consumption of wine rather than its prohibition. Their attitudes on moderation persisted into the fourth century AD supported by St. Paul and other fathers of the early Church. However, rather than castigating the consumption of wine, they emphasized its medicinal qualities as a gift of God. In the Central Middle Ages (500–1350 AD) castles, monasteries, cathedrals, and manors replaced cities as the basic units of social life

and became the centres of communal organization. Production of wine
and beer passed almost entirely into the hands of the Church. Since the
dawn of history, production of beer and wine featured prominently in
the economic structures of Western societies. Grain continued to be the
basic food of peasants and beer their normal beverage. Drinks of mead,
rustic beers, and wines based on wild and cultivated fruits, e.g. cider,
also became popular. All of them remained virtually the same as in
antiquity and were considered indispensable to life and living both as
thirst quenchers and sources of energy. The most vehement critics
of their consumption were not vociferous, as they could not cite any
alternatives.

The income derived from brewing and viticultural activities represented an important element in the finances and ceremonies of religious
establishments and their upkeep. For nearly 1300 years the Church operated almost all of the best and biggest vineyards with considerable profit,
a telling example of the economic impact of beverage alcohol and its far-
reaching effects on societal structure.

Concentration by Distillation

The end of the Middle Ages brought the return of cities as centres of
civilization, the expansion of trade and economies, the reform of religious and cultural life. Political and legal regulations generated new
social systems, including those due to the expansion of European hege-
mony in the New World, Africa, and Asia. The ensuing complexities
altered generalizations based on geography and other secular categoriza-
tions, and their correlations either with the consumption of alcohol or its
abuse generated novel dimensions and consequences. The motivation
and considerations of use and abuse of alcohol became much more
dependent on regional, political, religious, and moral perspectives, heav-
ily influenced, to be sure, by tradition. The expansion of distillation and
consequent concentration of alcohol into much more potent spirits of
low volume coined novel social patterns and created new interfaces
between peasants, merchants, clergy, and aristocracy. (Figure 3) shows
an example of an early still[6].

The introduction of distillation resulted in the first major change in
the mode and extent of human alcohol consumption nearly 10000 years
after the initial introduction of brewing and viticulture. Most importantly this initiated a transition from the consumption of beer and wine
as nutrients, liquid replacements and a source of calories to consump-

Fig. 3 An early distillation apparatus (from ref 6).

tion in amounts sufficient to cause pathology while focusing attention on the potential harm of alcohol. Distillation became the true benchmark of the negative effects of alcohol. Concurrently, there were persistent efforts to demonstrate its medicinal powers which generated trends to make its consumption a virtue from which to derive benefits that would counterbalance any possible deleterious consequences.

Efforts to recognize and attribute curative powers to alcohol preoccupied much of the medical thought of the times resulting in a balancing act of questionable merit or success. The monasteries of the Church perpetuated brewing and viticuture, the monks being the sole societal group with sufficient time, education, and patience to perfect and protect these skills while satisfying their own needs and demands as well as those of the Princes of the Church and governing classes for these products, to which spirits had now been added.

Distilleries of 'spirits' generated new areas of commerce through merchants and apothecaries who dealt with their production, distribution and sales. The new products eventually encompassed 'aqua vitae', brandy named after the Dutch brandewijn or burnt wine, liqueurs, whiskeys, gin and vodka. Their alcohol contents were considerably higher than those of beers and wines (Table 1).

A number of sources indicate that distillation was practised early at the Medical School of Salerno in the eleventh or twelfth centuries[6]. Salernus (d. 1167), a Salerno physician wrote a summary of alcohol, its pathology and therapeutics containng one of the earliest procedures for its distillation and praising it as a panacea in and of itself. Remarkably, its distillation from grapes and grains seems to have been invented and practised first by Arabian alchemists in Asia Minor whose skills and knowledge were transmitted to the University of Salerno in Italy whose scholars enjoyed close contacts with them and introduced their findings and practices into Western society. During this early period religious taboos and injunctions neither interfered with nor prohibited the production and utilization of alcoholic beverages in the Muslim world. This suggests—but does not prove—that initially Mohammed did not decree abstinence and the Koran did not contain explicit directions towards this end.

In the latter parts of the thirteenth and the early fourteenth centuries, knowledge of distillation spread from monastic to medical and

Table 1. Milestones of distillation

Date A.D.	Spirit
1200	Whiskey (Irish)
1250	Whiskey (Scotch)
1500	Vodka
1500	Benedictine
1550	Cognac
1600	Rum
1600	Chartreuse
1650	Gin
1700	Absinthe
1750	Drambruie

alchemistic circles, and the descriptions of some of the medical benefits of the products were almost poetic. The monk Ramond Lull (1235–1315), a missionary to the Muslim world, thought spirits to be marvellous medicaments for illnesses 'an emanation of the Divinity newly revealed to man, an element destined to revive the energies of modern decrepitude hidden from man since antiquity, because the human race was then too young'[5]. Arnald of Villanova, professor of medicine at the University of Montpellier in France established the use of wine as the basis of a system of therapy. He commented that it suits every age, time and region, is becoming to the old by alleviating their dryness and constituting food for the young whose nature it simulates. 'No physician of note blames healthy people for using it unless it is for the quantity of water with which it is mixed. It is most friendly to human nature and has miraculous, therapeutic properties'.

In 1527, Hieronymous Braunschweig (or Brunschwig) (d. 1533), an Alsatian physician, published a Book on Distillation[6] in London which included illustrations of a distilling apparatus along with an extensive discussion of the medicinal virtues of distilled spirits. (Fig. 3) Braunschweig states that

> 'Aqua vitae is commonly called the mistress of all medicines
> ... It causes a good colour in a person. It heals baldness and
> causes the hair well to grow, kills lice and fleas. It cures
> lethargy ... causes good digestion and appetite ... and takes
> away all belching and draws the wind out of the body. It eases
> ... the pain in the breasts when they be swollen, and heals all
> diseases in the bladder, and breaks the stone. It withdraws
> venom that has been taken in meat or drink. It heals all
> shrunken sinews, and causes them to become soft and right ...
> It heals the bites of a mad dog, and all stinking wounds, when
> they be washed therewith. It gives also courage in a person, and
> causes him to have a good memory'.

He also warned, however, that 'it is to be drunk by reason and measure ... five or six drops in the morning, fasting, with a spoonful of wine' (from the first English translation Braunschweig, 1527).

In assessing the role of alcohol and alcoholic beverages during the Dark Ages, the perpetual recurrence of The Black Death or Plague must be noted. Drinking of spirits followed very closely on the heels of the Black Death between 1348–1349 which, like previous ones, overshadowed all other considerations during the Middle Ages, including wars. No remedies of any kind were known and those that were tried proved useless against the plague whose source was unknown but which decimated the populations of Europe by as much as two-thirds in a given generation. The resultant devastation, grief, and hardship were of almost unimaginable magnitude even exceeding the consequences of wars. Nothing including alcohol proved to be effective.

Meanwhile knowledge of distillation spread at different times in different places, with Italy, Germany, and north Europe in the vanguard and both monasteries and apothecaries becoming large scale producers and distributors[5]. Table 1 summarizes the history of the principal distilled beverages giving approximate times of their introduction and popularization. In 1320, distilling became an industry in Venice, Italy. Venetians specialized in sweetened alcoholic beverages ('liqueurs') composed of alcohol, sugar or syrup, and flavouring agents. The techniques were introduced to and practised in France the same year and were propagated from there and became widespread.

The seventeenth century was a period of experimentation to improve the taste of spirits, but the quality of liqueurs was still less important than the quantity and cost. Stills were modified to increase their speed, and methods were sought to improve tastes of the products.

Beginning about 1450 economic recovery increased urbanization, made goods available and generated new standards of luxury contrasting with the prevalent abysmal poverty and class conflicts. An age of ostentation, gluttony, and inebriation followed. Negative effects of drunkenness were acknowledged, and governments tried to impose restrictions to control the results but to little avail. Intoxication interfered with the 'new order of rationality' reflecting loss of physical and psychic control at the expense of efficiency, time and order. The advocacy of abstinence coincided with the advent of the Reformation. Much as this has been inferred to imply a correlation, neither the Catholic Church nor leaders of the Reformation advocated the drinking of spirits nor did they oppose their use. There certainly wasn't any break between Protestantism and Catholicism on that basis. Both considered wine as one of God's creations meant for human enjoyment and benefit. Protestantism emphasized the separation of the 'secular from the sacred' but relied on civil rather than church authority to affect social behaviour based on moral sentiments instead of customs. The austere moral codes of the Hittites, Anabaptists, and Quakers exemplified the extremes of views emphasized by posterity but not by contemporaries, followed later on by German Pietism and British Methodism.

The call for temperance and even prohibition dates back to Hebrew, Greek, and Roman times and was reiterated throughout history. The reasons were as varied as the political systems, social structures and religious motivations which brought them about but all ordinances, regulations, laws and decrees of various forms as well as threats of punishment were ineffective in controlling alcohol consumption.

Potable boiled water (coffee, tea)

This synopsis has intentionally underemphasized major issues of public policy, public health, and scientific or medical knowledge. The introduction into Western Society of the boiling of water prior to its consumption in the form of coffee, tea, and cocoa brought about major changes in the West which made water potable (albeit unbeknown to the consumers) and free of poisonous (bacterial) side effects. In the nineteenth century, the purification of water and its separation from sewage through sanitary engineering eliminated noxious, toxic, and infectious agents ultimately replacing beer and wine as liquids requisite for human use. This totally altered the quantity of water consumed while eliminating alcoholic beverages for that purpose.

William Harvey (1578–1657) quickly recognized the value of substituting coffee for spirits and promoted drinking of coffee by praising its therapeutic benefits and considering it a panacea in and of itself and a cure for drunkenness, replacing ale, beer or wine; he urged his students to advocate it as an alternative for drunkenness. He was not alone. Among many others the Oxford Coffee Club and a coffee house at Wadham College, Oxford, was established, the latter becoming the birthplace of the Royal Society. Coffee houses and coffee drinking spread quickly throughout Britain, so that brewers found them competitive and demanded their taxation. Similarly tea and cocoa required the boiling of water, but among the resultant new liquids, coffee acquired the greatest number of adherents. Dr. Thomas Willis, a distinguished Oxford physician advised that patients should be sent to coffee houses rather than apothecaries, a move from which he expected marvellous benefits.

In London more coffee was consumed from 1680 to 1730 than in any other city in the world, perceptibly diminishing drunkenness among the upper English classes. Banking, shipping, insurance, stock, commerce, and trade were now conducted in coffee rather than ale houses paving the way for the future establishment of Exchanges. By 1700 coffee, tea, and chocolate had become the reigning luxury items of English cities.

In Prussia, coffee drinking spread from beginnings at the court to the higher circles of the bourgeoisie, and beer consumption declined during the reign of King Frederick II who had turned his attention from generalship to economics to bring about financial reconstruction of Prussia at the end of the Seven Years War in 1763. Both tobacco and coffee were major imports and interfered with his plans to establish a favourable trade balance with exports exceeding imports. In 1777, he issued a manifest dictating to his people, especially soldiers, to drink beer for its nourishing qualities:

'It is disgusting to notice the increase in the quantity of coffee used by my subjects, and the amount of money that goes out of the country as a consequence. Everybody is using coffee; this must be prevented. His Majesty was brought up on beer, and so were both his ancestors and officers. Many battles have been fought and won by soldiers nourished on beer, and the King does not believe that coffee-drinking soldiers can be relied upon to endure hardships in case of another war'.

Drinking coffee spread beyond the fashionable world, largely due to a greater availability and affordability of the commodity. While all other prices rose during the eighteenth century, a cup of coffee remained almost unchanged. From 1750 on, coffee consumption in France tripled. In 1657 the English East India Company first imported tea which was drunk very weak with sugar. On initiation of the British East India Company's direct trade with China in 1720, consumption in England rose considerably. While tea became fashionable, it remained prohibitively expensive for a long time, but in the course of the nineteenth century it became not only the usual beverage in the morning and evening but was generally consumed in large quantities at dinner.

Water purification and sanitation

Between 1801 and 1850, the population in the urban areas of Europe increased significantly. Industrialization concentrated ever increasing numbers of people in decreasing amounts of space. Thus, the population of Great Britain more than doubled and in Glasgow it even quadrupled. The concomitant hygienic problems focused on the separation of sewage from drinking water. Prior to 1900 sewage mixed with the water runoff into rivers and lakes and was the source of drinking water. The consequences propagated cholera, typhoid fever, and other bacterial infections referred to as 'miasmas' prior to the recognition of their bacterial origin which remained hidden until the relevant microorganisms were identified e.g. *Bacillus typhimurium* and *Bacillus vibrio cholera* in 1880 (Ebert) and 1893 (Koch), respectively.

Before the middle of the nineteenth century, little or no attention had been given to the quality and quantity of water available and consumed. Enlightened medical leaders and their students then focused on cleanliness, personal hygiene, and public health. The need for the construction and design of water supply and sewage removal systems and their separation failed to be recognized since the concurrent financial and engineering challenges were not self-evident. More than half a century passed before

Fig. 4 Louis Pasteur (1822–1895) who discovered fermentation, yeast, and bacteria. (From G.L. Geison, The Private Science of Louis Pasteur, Princeton University Press, Princeton, New Jersey, 1995.)

governments were persuaded to accept the importance and magnitude of the resultant public health problems as their responsibility and to devise financial and technical (engineering) means to solve them.

The perfection of the compound microscope and the recognition of bacteria and other microorganisms as the sources of the ill effects of water by Spellanzani and Pasteur, and the effects of sterilization, as well as Koch's cultivation of bacterial growth on solid gelatin media in Petri dishes made sanitary bacteriology the biological solution of the potable water problem (Fig. 4).

Between 1849 and 1854, much as they did not know the specific nature of the offensive agent(s) in water, two British engineers, Snow and York had already conclusively traced the source of a 'miasma' (a pseudonym for epidemics used prior to the discovery of bacteriology) in London and its deadly consequences to the effluent sewage of a specific house and the spillage of its contents into the Thames. Remarkably, Michael Faraday's advice on the pollution problem was sought and immortalized in a Punch

Fig. 5 Michael Faraday vainly introduces himself to Father Thames as the last resort in solving the pollution problem. Reproduced from *Punch*.

Cartoon which Professor Peter Day kindly made available from the archives of the Royal Institution (Fig. 5).

Relatively rapidly there ensued the widespread introduction of sand filters; bacteriology demonstrated their remarkable effectiveness in reducing the bacterial content of water.

Water safe for human consumption was neither known nor available but a century ago. The boiling of water for that express purpose was not practised in the Western world, much as making coffee and tea accomplished this end. With the aid of hindsight, the injunction of temperance groups to drink water instead of beer or wine as 'safe' beverages, without doubt well-meaning and constituting 'common sense' at the time, actually turned out to have been ill advised and even counterproductive. Until the turn of the last century, water safe for human consumption did not exist, and nearly half a century passed thereafter to make it reality.

Now, the suitability of a water supply safe for human consumption is established by chemical, physical, and biological analyses. Among those, microbiological examination is the most valuable as it detects and estimates the presence of pathogens. The existence of significant numbers of coliform bacteria in water is considered presumptive evidence of fecal contamination and the potential presence of pathogens. Their removal and/or destruction is pursued based on standards set by the World Health Organization.

The development of sanitation and purification of drinking water in Western industrialized countries before the end of the nineteenth century completely eliminated alcohol containing beverages as components of the human diet essential both to maintain body fuids and electrolytes, an achievement that is rarely identified or cited[7,8].

The recognition of alcohol-related diseases as a medical problem

At the end of the eighteenth century the largely spiritual views of both Methodists and Quakers regarding the pathological consequences of consuming distilled spirits had become an important aspect of medical thought and teaching in Great Britain, especially at the Edinburgh College of Medicine. There, two students, Benjamin Rush (1745–1813) (Fig. 6) and Thomas Trotter (1760–1831) (Fig. 7) made the first pacesetting contributions to the clinical recognition of the excessive use of alcohol and its consequences. Their reports are seemingly original and independent, even though they published their observations and conclusions almost contemporaneously. I have not come across records indicating that they knew each other or ever met. One may not even have been aware of the other's publications, which may not have been too surprising, given the period in which they worked and where they lived: Rush practised medicine in Philadelphia while Trotter was a medical officer in the British Navy.

In 1784 Rush published his views on spirits and their effects on drunkenness[9]. Rush believed that beer and wine were healthful when consumed in moderate amounts. He called chronic and persistent consumption of spirits and the ensuing drunkenness a distinct, severe disease leading to physical doom—and ultimately death. He pointed out that drinking of spirits leads to addiction which leaves the victim helpless to resist. In his view this made drunkenness not a vice or personal failing, as had been thought widely, but a disease, since alcohol controls the

Fig. 6 Benjamin Rush (1745–1813), who recognized the clinical consquences of prolonged consumption of alcohol. (From M.E. Lender, J.K. Martin, *Drinking in America*, A History, The Revised and Expanded Edition, The Free Press, New York, 1987.)

drinker, not the reverse. In Rush's view, once addicted to alcohol, 'even a saint would have a hard time controlling himself'. Before his death in 1813 he published a 'moral and physical thermometer' describing the progressive nature of alcohol addiction.

Thomas Trotter wrote an extensive essay on drunkenness as a medical thesis in 1788; it was published much later (1814) as a book in which he also described alcohol abuse as a disease (Fig. 7)[10]. He recognized and specified that habitual and prolonged consumption of alcoholic spirits caused liver disease accompanied by jaundice, wasting, 'epilepsy', and 'madness', manifesting at times when the afflicted patients are sober and not necessarily consuming alcohol. Trotter called drunkenness a disease of the mind. Both he and Rush concluded that the consumption of distilled spirits in quantity for long periods causes a serious, fatal disease. Their insight, remarkable for its clarity, was not accepted widely for almost another century.

Rush's thoughts regarding the consumption and consequences of hard liquor profoundly affected public and medical thought in America, and helped lay the foundation for many of the trends and viewpoints which ultimately led to prohibition a century later. Apparently, the religious

Fig. 7 Thomas Trotter (1760–1832); he described the clinical picture accompanying chronic alcohol ingestion (from ref. 10).

persuasions prevalent at Edinburgh College of Medicine played a significant role in the interests and motivations which brought both Rush and Trotter to the problem. Rush's views received a great deal more public attention in the colonies than did Trotter's in Britain. In part, this may have been due to the fact that Rush's religious, political, and social activism, combined with his prominence as a signer of the Declaration of Independence of the colonies, secured him a large and attentive audience. This clearly contrasted with Trotter's social setting. As a medical officer in the British Navy, he did not command a similarly large and devoted following. Subsequent studies in the nineteenth century were greatly reinforced by the emergence of structural pathology guided by Virchow in Germany, which ultimately described the structural changes which accompany excessive alcohol intake (Fig. 8). Jointly, these medical advances heralded the recognition and appreciation of what is now known to be one of the most important world health problems that affects millions of individuals in the USA alone.

Fig. 8 Rudolf Virchow (1821–1902) who recognized and described the microscopic and gross anatomical changes accompanying disease and accounting for the clinical manifestations. (From E.H. Ackerknecht, *Rudolf Virchow, Doctor, Statesman, Anthropologist*, University of Wisconsin Press, Madison, 1953.)

Chronic effects of alcohol consumption

Though most Americans are believed to drink or have drunk alcohol at one time or another in their lifetime, only one in twenty 'abuse' it, a relatively small proportion. Nevertheless, the death rates due to vascular disorders and cancer are now the only ones which exceed those due to alcohol-related diseases. The experimental study of the problem has been difficult, and has been plagued by psychological, social, religious, legal, and political biases.

'Alcohol dependence of the ethanol type', is the phrasing used by the WHO. It is worded diplomatically to avoid social conflict and is said to exist 'when consumption of ethanol by an individual exceeds the limits accepted by his culture and is sufficiently great to injure health or impair social relationships'. This includes closet drinking, inadequate job performance, blackouts, dependence on drinking to function normally, and inability to stop drinking. Loss of employment, occurrence of binges, violence, and legal and family conflicts are important manifestations of alcohol dependence.

Clinical features can either be primarily neuropsychiatric varying widely from pleasant conviviality to angry argumentation, and ultimately delirium tremens (DTs), vocal, visual or auditory hallucinations, delusions, paranoia or fear. Alternatively or simultaneously they can be gastro-intestinal, progressing from hyperacidity, indigestion, and bleeding, to stomach ulcers, and cancer (Table 2). Wernicke's and Korsakoff's diseases represent extreme complications and are the result of prolonged, severe vitamin B_1 depletion of the nervous system due to either diminished or inadequate vitamin intake and/or metabolism.

Alcohol can inflame the oesophagus and stomach accompanied by gastrointestinal bleeding and cause liver disease (cirrhosis) followed by jaundice and pleural or abdominal oedema. Acute and/or chronic alcoholic pancreatitis are important complications and account for more than 7% of all cases of that devastating abdominal catastrophe. Cancer is the second leading cause of death in alcohol abuse and the number of carcinomata of the pancreas, oesophagus, stomach, liver, and breasts exceed those of the general population ten-fold.

In contrast, modest doses of alcohol have been said to be beneficial for the cardiovascular system. One or two drinks per day over long periods are believed to decrease the risk of cardiovascular disease.

Table 2. Diseases associated with excessive alcohol ingestion

----

- Gastrointestinal, muscular and renal

  > hepatitis
  > cirrhosis, cancer
  > varicosities and bleeding
  > hemochromatosis
  > pancreatitis
  > muscular atrophy
  > kidney failure

- Neurological

  > central nervous system, cerebellum,
  >   peripheral nerves
  >   Wérnicke and Korsakoff's disease
  >   (thiamine deficiency)
  >   delirium tremens
  >   polyneuritis
  >   cerebellar atrophy

Heavy drinking during pregnancy transports both ethanol and acetaldehyde, the product of ethanol oxidation, across the placenta and interferes with normal fetal development.

The presence of ethanol in blood and other tissues confirms the existence of acute intoxication. There are, however, no reliable laboratory tests to establish chronic and prolonged ingestion of alcohol and possible 'addiction' to it retrospectively.

Biology, genetics, pathology, and therapy

The enzymatic basis of alcohol oxidation

I will not here enter into the details of the biochemistry of alcohol metabolism, which has been studied intensively for more than 60 years and has remained in the forefront of modern biochemistry. Alcohol dehydrogenase is a cytoplasmic enzyme found primarily in the liver which oxidizes most of the ingested alcohol and is thought to be the major route for its detoxification. At least two additional metabolic pathways for the oxidation of ethanol are known but are quantitatively less significant. None of them exist in brain or other neural tissues. An enzyme cofactor (NAD^+), containing the vitamin nicotinic acid, is essential for the oxidation of alcohol to acetaldehyde.

The affinity of alcohol dehydrogenase for ethanol is very high so that the enzyme is saturated with it even when the levels of alcohol in the body are quite low, an important fact since no other ethanol receptors are known. At least six classes of alcohol dehydrogenases occurring in the cytoplasm of most human tissues oxidize ethanol. Those and a series of aldehyde dehydrogenases have been studied intensively. Aldehyde dehydrogenases oxidize the resultant acetaldehyde to acetate which in turn is metabolized to water and carbon dioxide.

Table 3. The alcohol dehydrogenase—aldehyde dehydrogenase pathway

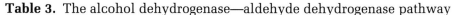

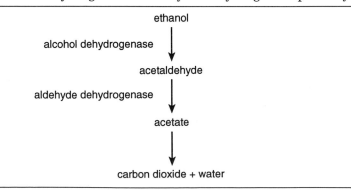

The brain apparently does not oxidize ethanol ingested or produced in the body; therefore this organ seems to play an insignificant role in the utilization and disposal of ethanol. There is no known animal model which successfully mimics alcoholic cirrhosis; this is hindering efforts to understand the cause of this disease and so to develop effective therapy.

The clinical features and pathological consequences of prolonged and persistent alcohol consumption and the delineation of the biochemical pathways of its oxidation have been the basis for the identification of the pathway which detoxifies alcohol as well as the design of therapeutic agents to treat alcohol abuse.

Genetics and the alcohol dependence problem

Genetic studies of the use and abuse of alcohol have been emphasized increasingly in both experimental and intellectual approaches to the problem. Genetic views of the alcohol problem have been favoured by a variety of clinical observations, the results of which have proven both persuasive and provocative to many. They have found many adherents and supporters who have generated a great deal of both biochemical and clinical work. Among these are observations on twins, adopted children of alcohol-abusing parents and their progeny, follow-up in several generations, and observations of such groups in different countries.

Judging by family, twin, and adoption studies, alcoholism is a multi-factorial disease that arises from a combination of biological and genetic factors which have stubbornly resisted definitive analysis. For close male relatives of primary alcoholics, the risk of developing the disorder is approximately four-fold greater than for controls. Genetic population studies reveal that the risk of male offspring of alcoholics becoming alcoholic is four-fold greater when such individuals were adopted at birth and raised without knowledge of their own history or of that of their biological parents. The evidence supporting genetic influences in alcoholism has stimulated numerous studies of children of alcoholics and of the biochemistry which may underlie it.

A different line of work has centred on alcohol-dependent rodent colonies, particularly rats, selected, bred, and reared for genetic traits along with efforts to single out potential contributing or facilitating features such as taste of food and psychological conditioning of the animals. It is thought that through this approach drug therapy for the treatment of the human problem might be identified. The hope for significant extension of effective therapeutic agents by this means has not materialized thus far. The possibility that inheritance and genetic endowment play roles in the processes involved has been explored, and some investigators have reported on the existence of predisposing genes. The interpretation of the abundance of data has led both to controversial claims and their refutation. Remarkably none of these genetic studies have generated insight that is accepted universally—with one important and thus far unique exception.

In approximately half of the Chinese and Japanese and related Oriental populations e.g. Indonesians, Koreans, and Vietnamese, the metabolism of ethanol genetically differs significantly from that of virtually all Occidentals. In these Oriental populations, a genetic variant results in a non-functional mitochondrial aldehyde dehydrogenase causing the 'flushing syndrome'. These individuals are intolerant to alcohol consumption which makes them acutely ill. As a consequence, these numerically substantial population groups avoid alcohol consumption altogether. As would be expected, the number of individuals who develop subsequent problems with alcohol consumption is substantially lower than would be expected in a comparable Occidental group. Hence, chronic alcohol abuse is quite infrequent in those Oriental populations, and the incidence of alcoholism, e.g. in Japan, is considerably lower than in western countries.

The characteristic symptoms of the ensuing alcohol sensitivity include facial flushing, increased skin temperature, nausea, vomiting, abdominal discomfort, dizziness, headaches, increased heart rate, and fainting. These symptoms correlate with an accumulation of acetaldehyde (which is not oxidized quickly and effectively) in the blood and tissues of these individuals. Jointly, these facts suggest that individuals who lack, or are deficient in the mitochondrial aldehyde dehydrogenase are physiologically protected against 'alcoholism'.

The lack of a functional mitochondrial aldehyde dehydrogenase which results in the avoidance of alcohol intake is the only specific human genetic alteration known thus far. Its very existence has fuelled interest in and directed attention to genetics and its manifestations as an approach to the conundrum of the use and abuse of alcohol.

In these individuals, the reaction to alcohol is predictable and sufficiently unpleasant so that the majority of those who have inherited the null mitochondrial aldehyde dehydrogenase trait avoid alcohol. In a sense, this then protects them from the pathological consequences of alcohol excess. However, attempts to mimic this through the use of drugs like antabuse have not been encouraging, despite the fact that they have also focused attention on possible alternative measures employed in the Orient, particularly China, to cope with problems of alcohol abuse, seen in that part of the Oriental population that does not exhibit the above genetic mutations.

Pharmacology and therapeutics of alcohol abuse based on genetics

Antabuse was the first agent suggested for the therapy of alcohol abuse in 1948, now more than forty years ago[11]. Its suitability and effectiveness have been examined but found disappointingly inadequate; the drug may even be harmful. Aldehyde dehydrogenases have been detected, isolated, and studied both in cytoplasm and mitochondria. Antabuse inhibits the mitochondrial and cytosolic aldehyde dehydrogenases[12]. When acetaldehyde produced by alcohol oxidation accumulates in the body it causes intense flushing, nausea, dizziness, sweating, throbbing headaches, fainting, fall of blood pressure, and ultimately shock.[13] These symptoms are analogous to those experienced by Orientals who are known to lack mitochondrial aldehyde dehydrogenase on a genetic basis (see above). Antabuse treatment has been dubbed 'aversive' but is more properly described as 'deterrent': its consequences are so unpleasant that many patients given this drug may, indeed, cease drinking alcohol rather than experience the reaction to it. Alternatively, of course, such patients may continue to drink but not take the drug, considering that the consequences of taking it are worse than having the alcohol problem, i.e. 'The cure is worse than the disease'. This therapy has not been accepted universally and remains of questionable value. Moreover, extensive clinical evaluations are virtually unanimous in the conclusion that antabuse and its analogues are not effective therapeutically. A recent authoritative review[13] states that well controlled studies have never demonstrated that the administration of antabuse has resulted in sustained abstinence of drinkers or is more effective than placebos[14].

It is discouraging that the search for other agents with therapeutic potential, based on the known oxidative metabolism of ethanol, have not been successful thus far. Very recently I have studied the antidipsotropic action (i.e. ability to cause aversion to alcohol) of a group of isoflavones, among them daidzin and daidzein, which I have found in the kudzu plant (Fig. 9) and which are also known to occur in legumes such as soybeans. They significantly inhibit the desire of Syrian golden hamsters to drink alcohol, which this species prefers to water when given a free choice. These isoflavones are particularly abundant in the roots of the kudzu plant, an extract of which has been employed in China as a tea for nearly a thousand years to reduce alcohol intake. My studies verify the favourable reports in the Chinese herbal medical literature, and we are currently in the process of exploring their potential. It is premature to speculate on the effect of these chemical compounds in humans at this time[15,16]. Daidzin and daidzein (Fig. 10) inhibit mitochondrial aldehyde dehydrogenase and concomitantly decrease the desire of both hamsters and rats to drink alcohol, apparently without adverse effects (Fig. 11). Daidzin does not affect overall acetaldehyde metabolism in golden hamsters but inhibits acetaldehyde oxidation catalyzed by isolated hamster liver mitochondria.[17] Present evidences indicate that daidzin may suppress ethanol intake by modulating the metabolism of a physiological aldehyde other than acetaldehyde. Details of the current investigation and their importance to ethanol use and abuse will be published shortly.[18]

The mesolimbic reward system

The diligent search for promising alternative pharmacological and therapeutic approaches to alcohol abuse has been extensively documented in a Symposium of the Nobel Foundation[19]. The study of the mode of action of addictive drugs has placed significant emphasis on a hypothesis which proposes a mesolimbic reward system whose existence has not been proven experimentally. Receptors of GABA, dopamine, glutamate, serotonin, opium, heroin, cocaine, and corticotropin-releasing factor, as well as receptors of other neurotransmitters continue to be identified in large numbers and have been show to play significant roles in the mode of action of addictive drugs, such as heroin. Both their individual and collective role(s) in the mechanisms of action and pharmacology of addictive drugs have been and remain under expanding, intensive investigation. The speculation that ethanol acts in a manner similar or identical to such drugs has received such wide attention that many now consider this a fact. There is, however, no direct experimental evidence for this assumption[20], a disappointment which has both caused much discouragement as well as persistent efforts to re-examine and extend present knowledge.

Naltrexone is one of the drugs which effectively antagonizes the binding of heroin to its receptor; it has proven effective in the treatment of heroin addiction and has now also received wide attention for the treatment of alcohol abuse[21]. Both naltrexone and acamposate[22] have been examined for their antidipsotropic effects and proposed as possible candidates for the treatment of alcohol abuse based on the underlying premise that their mechanism of action for heroin would resemble that for alcohol, even though there is no experimental evidence for that speculation. No alcohol receptor has been demonstrated that would support this conjecture and the use of these agents in the treatment of alcohol abuse. In spite of the lack of experimental data justifying the speculation these drugs have now undergone widely publicized clinical trials for the treatment of alcohol abuse, but their effectiveness is in doubt. Together

藥材

葛花

葛根
Pueraria pseudo-hirsuta Tang et Wang

Fig. 9 The kudzu plant.

with psychotherapy, they do seem to facilitate abstinence for longer periods but the evaluation of long-term success independent of other measures remains to be established. Both the effectiveness and mechanisms of action must be ascertained as must the validity of the preliminary clinical findings and of the analogies drawn.

The equivalence of addiction to alcohol and to heroin or cocaine remains to be documented. Nevertheless, both the circumstances that have led to the use of naltrexone and acompensate and the interpretation of the resultant data as well as the effectiveness of the agents must receive close inspection regardless of the validity of the rationale.

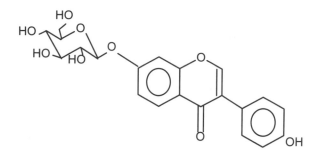

Fig. 10 Structure of the isoflavone daidzin.

Reports of the studies of naltrexone emphasize a high failure rate when used alone. It has been effective only as part of treatment programmes that combined its use with psychotherapy. Nevertheless, these studies suggest a novel approach to treatment; yet other combinations of medications and psychotherapy could prove useful empirically. The observed synergism of drugs and psychological measures could hold hope for the extension of joint interventions.

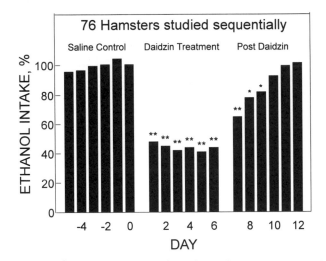

Fig. 11 Daidzin suppresses ethanol intake in Syrian golden hamsters: saline (1 ml/day) and daidzin (150 mg/day) were injected interperitoneally. The mean of ethanol intake measured during the saline control period is taken as 100%. Values are the means of measurements obtained from 76 hamsters. Ethanol intake measured on each day during and after daidzin treatment period is compared with that measured on the day before daidzin treatment began (Day 0). ** $P <$ 0.001, * $P < 0.01$.

Turning to yet other possible approaches, the role of 'appetite' in alcohol consumption has not been addressed explicitly in attempts to control consumption. At this juncture the biochemistry of appetite is largely unknown, whether appetite for food, liquids, or pleasure in general. Clearly, the relevant molecular biology must be investigated if a rational therapy is to be designed on that basis. While an approach to the therapy of alcohol abuse by appetite modification remains to be defined, it might be possible to achieve such aims empirically as exemplified by the Chinese experiences with herbal medicine cited.

It may be conjectured that the action of alcohol is mediated by multiple and variable neurochemical systems. The exact reinforcing effects of alcohol are speculative. There are intense efforts to extend and translate behavioural studies of the effects of alcohol to a neurobiochemical basis. In point of fact, the cause of alcohol abuse is unknown. Postulated causes include:

Table 4. Postulated causes of alcohol abuse—the result of different disciplinary views and orientations

---
sociology: historical realities; cultural attitudes
psychiatry and psychology: alcoholic personality; addiction; appetite
toxicology: dietary contaminants and imbalances
biology: genetics
biochemistry: metabolic pathways and mediators

Conclusions

A balance sheet of the current state of knowledge regarding the recognition, understanding, and treatment of alcohol abuse should include the results of medical, psychiatric, and therapeutic efforts and those of the great prohibition.

Prohibition of alcohol in the United States was a truly extraordinary experiment in public health on a scale never attempted before nor likely to be repeated soon. Unfortunately, the design of the experiment did not provide for suitable and valid controls to guide future medical and social actions. The resultant information is of variable quality, but does demonstrate a decline in long-term alcohol-related medical problems and of the national death rate due to alcohol-related diseases. Ultimately prohibition did alter the amount and characteristics of alcoholic beverages consumed; these drinks have, at least in part, been replaced by non-alcoholic fruit juices, soft drinks, and water.

Remarkably, the causes, diagnosis and treatment of alcohol abuse remain elusive and the long-term aspects of this affliction are among the major unsolved problems of chronic diseases such as cancer, athero-sclerosis, arthritis, AIDS, Alzheimer's disease, and ageing, for which there are no specific therapies based on known causes. Hence, therapeutic approaches for all of them, including alcohol, continue to be empirical.

So far, total abstinence is the only reliable treatment with predictable results. Little can be added to the vocal arguments for or against any par-ticular regime advanced throughout the last and this century regardless of their motivations, since none of them are directed at known targets; as a consequence, the results with all of them have remained imprecise.

Studies of the cell and molecular biology or biochemistry of 'addictive' drugs have not as yet given specific new insight that would serve as the basis for novel therapeutic measures regarding the abuse of ethanol. What-ever has been learnt in regard to addiction of drugs in general has failed so far to provide a basis for therapy of alcohol abuse in particular[17].

Pharmacological intervention with specific agents including antabuse has been uniformly disappointing so far. Multiple other agents that have been examined for their therapeutic suitability have also been found wanting. None of the drugs cited in the relevant literature have proven reliable and broadly effective therapeutically.

The anatomic and metabolic changes accompanying alcohol abuse are now well known, but the mechanisms that bring them about clearly are not. There are not even speculative hypotheses regarding possible mech-anism(s) of metabolic bases which are backed by valid experimental data that could serve as a guide to rational experimentation. The isoflavones that have been found to be effective in rodents remain to be studied in humans.

While progress in the diagnosis, understanding, and treatment of alcohol-dependency has been excruciatingly slow, public reaction to the problem has truly made headway. Society now appreciates the import-ance of gaining better understanding of the relevant biology, physiology, and pharmacology and leaves the experimental study and interpretation as well as application of the results to professionals. In 1918 Lord D'Abernon tellingly summarized and deplored the bias of society against a detached and unemotional discussion and view of the causes of alcoholism: 'Those who would give any attention to scientific work on alcohol do so less to gain knowledge than to find aims and arguments to support their preconceived opinion'[23]. It is comforting to note that this once prevalent stance seems to have been modified. At least the world is now receptive and listens. We hope that new experimental and intellectual avenues will give directions to the solution of the remaining problem.

Acknowledgement

The initiative and generosity of Edgar M. Bronfman, Sr, set in motion this historical research and enquiry into the basis of the use and abuse of alcohol.

References

1. J. Marcuse, *Diätetik im Alterthum. Eine Historische Studie*, Ferdinand Enke, Stuttgart, 1899, pp. 46–9.
2. M. Allen, *Chemistry in Britain*, 1996, **32**(5), 35.
3. W.J. Darby, P. Ghalioungui, and L. Grivetti, *Food: The Gift of Osiris*, Vols 1 and 2, Academic Press, New York, 1977.
4. S. Lucia, *A History of Wine as Therapy*, Lippincott, Philadelphia, 1963.
5. H. Braunschweig, *Kleines Destilierbüch*, London, 1527.
6. R.J. Forbes, *Short History of the Art of Distillation*, E.J. Brill, Leiden, 1970.
7. R.M. Sterrit and J.N. Lester, *Microbiology for Environmental and Public Health Engineers*, E. and F.N. Spon Ltd, London, New York, 1988.
8. J. von Simson, *Tech. Hist. Assoc. Ger. Eng.*, 1983, **39**, 4–9.
9. B. Rush, *An Inquiry into the Side-Effects of Ardent Spirits Upon the Human Mind and Body, With an Account of the Means of Preventing and of the Remedies for Curing Them*, 8th edn, Brookfield, MA., 1814.
10. T. Trotter, *Drunkenness*, Bradford and Read, Boston, and A. Finley, Philadelphia, 1813.
11. J. Hald, E. Jacobsen, and V. Larsen, *Acta Pharmacol. Toxico.* 1952, **4**, 285.
12. H.W. Goedde and D.P. Agarwal, *Pharmacol. Ther.*, 1990, **45**, 345.
13. P. Banys, *J. Psychoactive Drugs*, 1980, **20**, 243–60.
14. T. Tennant, *J. Am. Med. Assoc.*, 1986, **256** (11), 1489.
15. W.-M. Keung and B.L. Vallee, *Proc. Natl Acad. Sci. USA*, 1993, **90**, 1247.
16. W.-M. Keung and B.L. Vallee, *Proc. Natl Acad. Sci. USA*, 1993, **90**, 10008.
17. W.-M. Keung, O. Lazo, L. Kunze, and B.L. Vallee, *Proc. Natl Acad. Sci. USA*, 1995, **92**, 8990–93.
18. W.-M. Keung, A. Klyosov, and B.L. Vallee, *Proc. Natl Acad. Sci. USA*, 1997, in press.
19. B. Jansson, H. Jörnvall, U. Rydberg, L. Terenius, and B.L. Vallee, *Toward a Molecular Basis of Alcohol Use*, Birkhäuser Verlag, Basel, 1993.
20. A. Goldstein, *Addiction*, W.H. Freeman and Co., New York, 1994.
21. J.R. Volpicelli, A.I. Alterman, M. Hayashida, and C.P. O'Brien, *Gen. Psychiatry*, 1942, **49**, 876–80.
22. J.P. Lhuintre, N. Moore, G. Train, L. Steru, S. Langrenon, M. Daoust, P. Parot, P. Ladure, C. Libert, F. Boismare, and B. Hillemand, *Alcohol and Alcoholism*, 1990, **25**, 613–22.
23. L. D'Aberon, *Alcohol: its action on the human organism*, British Medical Council, 1918.

BERT VALLEE

Born 1919, obtained a degree in Medicine from New York University and subsequently held positions at MIT and Harvard Medical School in Medicine and Biochemistry. Since 1948 he has been a member of the Medical and Science Faculty of Harvard University where he has held various professional positions. He is an Honorary Professor of Tsinghua University, Beijing, China and Stellenbosch University, Stellenbosch, South Africa. He is a Fellow of the National Academy of Sciences of the USA and the Royal Danish Academy, and holds Honorary Degrees from the Karolinska Institute, Stockholm, the University Degli Studi Di Napoli Federico II, Naples, and the Ludwig Maximillian University, Munich. He was awarded the Order Andres Bello First Class of the Republic of Venezuela, and the Gibbs, Linderstrøm Lang, and Rose medals. He is the author of more than 600 scientific publications including books.

Incremental decisions in a complex world

CHRIS ELLIOTT

Introduction

In 1962 President Kennedy set the American nation the challenge of 'putting a man on the Moon and bringing him back safely before the end of the decade'. In 1969 Neil Armstrong said (or should have said) 'that's one small step for a man but a giant leap for mankind'. Between those two speeches tens of thousands of people made millions of individual decisions. One of them said 'I don't know how all of this mission works, I don't even know how most of it works but I do know how my bit of it works and it ain't gonna fail because of me'.

This article is about taking decisions. There are two problems—how do you take the individual decision that faces you when you do not have sufficient time and resources and how do you ensure that your decision fits together with all of the decisions that your colleagues are taking? The second problem is one of systems so I had better start by defining a system: a system is a set of components which, when they are brought together, exhibit properties that were not present in the components alone. These properties are called emergent properties and they often make it difficult to predict how a system will behave by examining the components alone.

I shall be coming back to systems later but let me illustrate an emergent property in action. Figure 1 shows a spring hanging from a support and joined by a short string to a second spring. A weight is hanging from the lower spring and loose strings are tied from the support to the top of the lower spring and from the bottom of the upper spring to the top of the weight. The question is: 'what will happen if I cut the short string joining the two springs?'. Clearly the weight will fall, but how far? Three possible options are shown in Fig. 2.

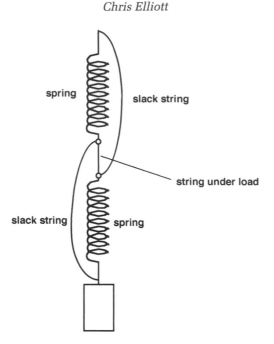

Fig. 1 The arrangement of springs.

In fact, when the string is cut, the weight does not fall but **rises** by around 10 cm. On the face of it, this is remarkable since a string under tension has been cut and the two things that were pulling it apart become closer together. However, if we redraw the picture as shown in Fig. 3, we can see the reason why. Electrical engineers will see it immediately—the springs were 'in series' before and are 'in parallel' afterwards. When they are in series, each carries the full weight. When they are in parallel, each only has to carry half the weight and so shortens.

In Fig. 3 I have included some illustrative numbers for the dimensions and a simple Hooke's law definition of the spring stiffness. There is no doubt—the weight rises. In fact, the behaviour is quite complicated and depends on the length of the strings, the mass of the weight, and the stiffness of the springs. If we increase the mass, so that the springs stretch to take up the slack in the 'loose' strings, the weight still rises (Fig. 4).

I shall come back to the properties of systems later. Before that I want to look more generally at the world of engineering design. One of the greatest engineers of the twentieth century, Sir Ove Arup, said when delivering the Institution of Structural Engineers Maitland Lecture in 1968:

> As in art, its problems are under-defined, there are many solutions, good, bad and indifferent. The art is, by a synthesis of

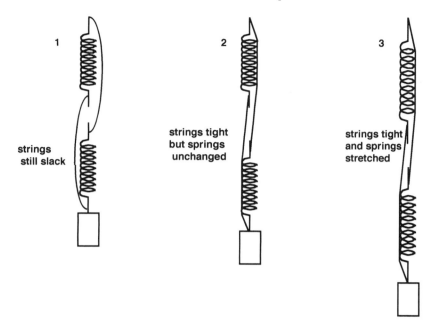

Fig. 2 Three possible options for the height of the weight after cutting the string. Position 2 is the point at which the two loose strings are just tight and the springs are unaltered. Position 1 is higher than position 2. Position 3 is lower than position 2 and requires the springs to stretch when the string is cut. I think that position 2 looks the most likely result, but not everyone agrees—in the Discourse upon which this article is based, the majority of the audience voted for position 3.

ends and means, to arrive at a good solution. This is a creative activity, involving imagination, intuition and deliberate choice, for the possible solutions often vary in ways which cannot be compared by quantitative methods.

He was talking about engineering design and that is my main theme but the words that he used might equally be used for problems in politics, business, or one's personal life. The engineer's dilemma can be illustrated easily—shall I use a 6 mm bolt or an 8 mm bolt? The former is lighter and cheaper but the latter is stronger and hence safer. There is no right answer, just as there is no right answer to the political decision on how much to spend on health, the business decision as to whether to invest in a new factory, or the personal decision on where to take a holiday.

I shall start by considering what I call 'simple' decisions—those are the ones where you know what it is you are trying to achieve and the difficulty is achieving it (I have put the word 'simple' in quotes because

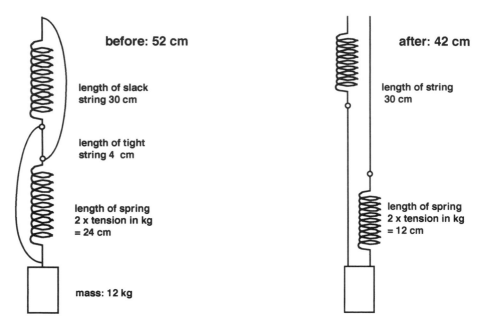

before: 52 cm

length of slack string 30 cm

length of tight string 4 cm

length of spring 2 x tension in kg = 24 cm

mass: 12 kg

after: 42 cm

length of string 30 cm

length of spring 2 x tension in kg = 12 cm

Fig. 3 The springs before and after cutting.

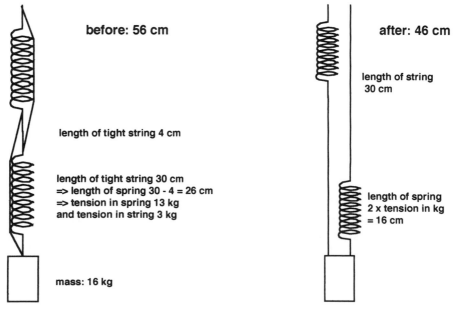

before: 56 cm

length of tight string 4 cm

length of tight string 30 cm
⇒ length of spring 30 - 4 = 26 cm
⇒ tension in spring 13 kg
and tension in string 3 kg

mass: 16 kg

after: 46 cm

length of string 30 cm

length of spring 2 x tension in kg = 16 cm

Fig. 4 The springs with a heavier weight.

these decisions are often far from simple in practice). After that, I shall consider complicated decisions, where you are not even sure what a good solution looks like. Finally I shall consider complex decisions which involve systems and their emergent properties.

Simple decisions

Investment decisions

Let's start with a decision-making problem that interests anyone who is paying into a pension fund—which investments should be made to max-imize the value of the fund. Let us assume that ten times per year the manager invests five per cent of the fund in a speculative investment. We will also assume that half of speculative ventures rise in value by ten per cent more than inflation. The other half fall in value by the same amount.

Clearly, if the fund manager selects his investments at random, with-out using his expertise in the markets, the average value of this fund exactly tracks the inflation index (if we ignore dealing costs). What would we expect to be the distribution of values of a large number of funds, each of which was managed according to this principle? After five years, the standard deviation will be $\sqrt{50} \times 5\% \times 10\% \approx 3.5\%$. We would therefore expect the value of the funds to follow the normal distribution curve as in Fig. 5, centred around the rate of inflation.

percentage growth per year relative to inflation

Fig. 5 Annual growth rate of funds using random investment decisions.

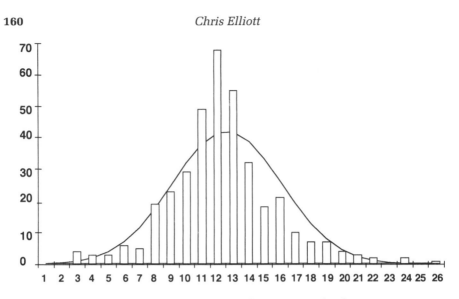

Fig. 6 Growth rate of actual investment funds.

What happens in practice? I used the Internet to find the growth rate of the 371 US growth funds that have been trading for five years. Figure 6 shows the histogram of those values, superimposed on the normal distribution of Fig. 5. The mean growth rate is 12.3%, slightly less than the growth in the Dow-Jones index over the same period which was 12.6%. The first observation is that the specialist fund managers have, on average, added no value by their efforts. They have just made enough to cover their dealing costs and fees. It is then interesting to note that more funds are grouped around the centre than the normal distribution would predict. This, together with the relatively small number of funds in the 'shoulders' of the distribution, indicates that many of the fund managers did not bother to trade as much as would be expected. Some funds did much better than would be expected—perhaps they are the insider traders. Finally, some did much worse than would be predicted—perhaps they are incompetent inside traders.

What does this tell us? It is not just a chance for an engineer to make fun of City dealers—the point is that there are some decisions that do not merit a great deal of effort. Where we do not have enough information to take a rational decision, or where the consequences of our choice are not predictable, there is no point in putting a great deal of effort into the decision. This is a very important message for an engineering designer (or any other decision maker). Since you rarely have enough time to analyse all of the issues that are amenable to analysis, don't waste time on those that are not.

Detection—two ways of getting it wrong

Countermatch is a computer developed by AEA Technology to recognize handwritten signatures. It works by calculating a score for the similarity between the signature and a reference version stored in its memory. If the score is above a threshold value, the computer says that the person making the signature is who he claims to be. If the score is below the threshold, it says that he is an impostor. The way in which we write our signatures varies so the system cannot be perfect—the score it will attribute varies around the correct value.

'Countermatch' has been successfully used by the Employment Service, the Ministry of Defence, and banks. Each of them had to take a difficult decision—at what level do I set the threshold? The Employment Service is concerned not to reject any valid applicant since this would be very embarrassing. It therefore sets the threshold quite low to avoid any false rejections. The Ministry of Defence, on the other hand, is concerned that no-one should gain unauthorized access to sensitive information and sets the threshold very high. It is prepared to annoy a few genuine users to be safe. The banks are even more subtle. If you want to withdraw £10, they set the threshold at a low value since it is better to make a few incorrect payments than to annoy customers. If on the other hand you try to withdraw £10 000, the threshold is set at a high level to ensure that they are paying the correct person.

It follows that there are two kinds of mistake:

- falsely accepting an incorrect signature
- falsely rejecting a valid signature

and the user can bias the decision-making system so as to make one less likely at the expense of increasing the probability of the other. This allows us to introduce the precautionary principle in decision making. You can estimate the consequences of each kind of mistake and bias the decision making process towards the kind of mistake that is less damaging.

This kind of decision making was first subject to rigorous mathematical analysis by the early developers of radar who called it 'detection theory'. Radar works by transmitting a powerful pulse of radio energy in a narrow beam. The pulse is scattered by any obstruction that it encounters and some of that scattered energy reaches the receiver of the radar set. The strength of that 'echo' is an indication of the size of the obstruction and the time between sending the pulse and receiving the echo is proportional to its range. A further complication is that the further away the obstruction, the weaker the echo.

Now let us imagine the radar on a warship. The pulse is transmitted

Chris Elliott

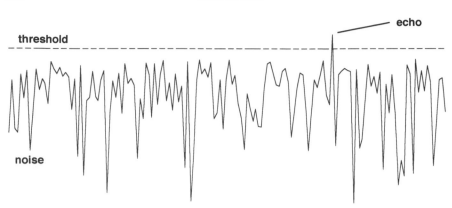

Fig. 7 A radar signal with a strong echo.

over the sea and is scattered by the waves and any other obstruction, be it a seagull or an Exocet missile. The challenge is to decide when the received energy is an echo from an Exocet, and to decide in time to take evasive action.

When the Exocet is near, the signal is very strong. If we plot the received energy over time, it might look like Fig. 7. The normal signal, labelled 'noise', is a combination of the echoes off waves and other benign obstructions and the inherent electronic noise of the receiver. We can set a threshold that will reliably detect the echo and hence know that the missile is present. If, however, we want to detect it at greater range, the threshold will have to be lower and we will start to get signals crossing the threshold which are not missile echos. Once again, we have the choice of two sorts of error (called false alarms and missed detections).

Where do we set the threshold? If it is too high we will not detect a missile in time to react. If it is too low, we will use up all of the ammunition in our Close-In Weapon System before a real missile is ever encountered.

False alarms are like the boy who cried wolf: they devalue the subsequent real detections. Decision makers, being mindful of the precautionary principle, have to weigh up the probability and consequences of each kind of error when setting the threshold.

Trends and fluctuations

We can also apply the ideas of decision theory to slowly varying phenomena, where the challenge is sometimes that of 'seeing the wood for the trees'. Let's take global climate change as an example. The change is slow and the fluctuations from year to year can mask the underlying trend. One of the techniques used by climate scientists is to look at historical data over a long period.

There are many sets of data that one could use. An example is the average temperature in December in England. Figure 8(a) shows that temperature every ten years since 1683. It is hard to detect any pattern or systematic change. If we start to calculate running averages, a trends begins to emerge. Figure 8(b) shows the running average over fifty years, superimposed on the raw data. The increase over the last ~100 years is beginning to emerge. Figure 8(c) shows two more running averages of increasing length. The solid line shows a clear trend, with a warm period from around 1750 to 1850, a cooling in the late Victorian times and a warming during this century. The dotted line shows that too much averaging smooths not only the fluctuations but also the underlying phenomenon that we are trying to detect.

Now the question is whether the trend that we see in this century is a cyclic variation or the beginning of a more serious effect. Clearly much more data is available but again we have to use the precautionary principle. If we decide that it does indicate a real problem and we are wrong, we will waste money and effort on changing the way we generate energy and run society. If we decide that there is no problem and we are wrong, we leave an even greater problem for our grandchildren.

Conclusion

This section of this article concerned 'simple' decisions. I hope that it is clear that actually taking the decision is often far from simple but at least we know what it is we are trying to do—detect the impostor, detect the missile, or detect climate change. Even that is not clear for complicated decisions.

Complicated decisions

Optimization—what is a good solution?

Let's take a simple optimization problem—the power produced by a car engine as the ignition timing is varied. There is a setting of the timing which gives the greatest power and any change, either advance or retard, reduces the power. When it is correctly set, a racing car engine produces much more power than a road car engine but the performance peak is much narrower (see Fig. 9).

Which is better? If you want to win a race, you have to have the power of a racing car engine at its peak. If you do not have that power, one of

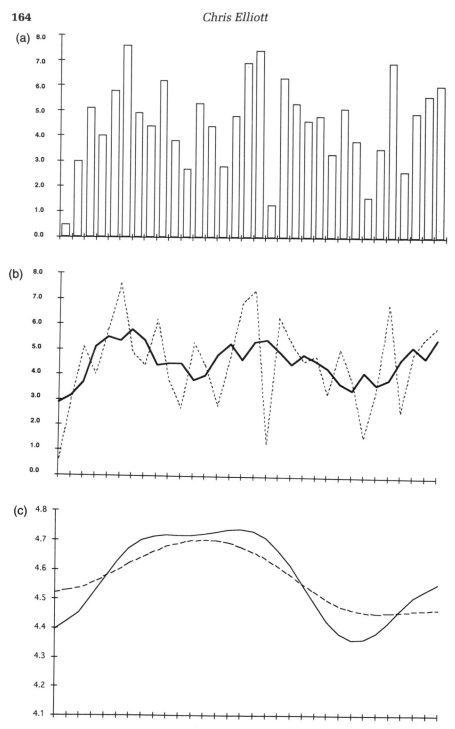

Fig. 8 (a) Average December temperature in England every ten years from 1683 to 1993. (b) The running average temperature over 50 years. (c) Further average of the temperature.

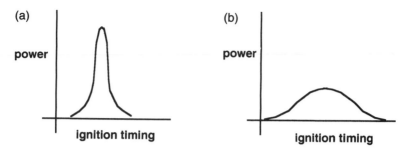

Fig. 9 Engine power as a function of ignition timing, (a) in a racing car and (b) in a road car.

the other competitors who does will beat you. However, the setting is very sensitive and any slight change will drastically cut the power available. If on the other hand you want to drive to work, reliability is more important than peak power and the road car performance curve is much safer. The answer to the question 'which is better?' is 'it depends on what you want to do'.

Let me give another transport example: most people would say that a thoroughbred race horse is better than a camel, but is that true when you are halfway across the desert?—more about camels later. First it is necessary to look in more detail at 'measures of goodness'. It is necessary to introduce two definitions:

- efficiency: how well does the solution use resources?
- effectiveness: how well does the solution meet the needs?

Consider a train pulling a load up a hill. The measure of effectiveness is the rate at which it raises cargo up the hill, which is the rate of climbing multiplied by the mass of cargo. What fraction of the total mass of the train should be allocated to the engine? It is reasonable to assume that the power of a train engine is proportional to its mass. The rate at which it can lift mass up a hill is proportional to the power of the engine divided by the mass of the train, which equals the engine fraction. The rate of raising cargo is the rate of climbing multiplied by the mass of cargo, and thus is proportional to the engine fraction multiplied by (1 − engine fraction). Figure 10 shows this measure of effectiveness plotted as a function of engine fraction. It is clear that the train is most **effective** when it is only fifty per cent **efficient**, that is, where half of the total mass of the train is engine.

A better known example of exactly the same effect is the electrical circuit shown in Fig. 11. If the aim is to heat the water as quickly as possible, the resistance of the heater, r, should be equal to the internal

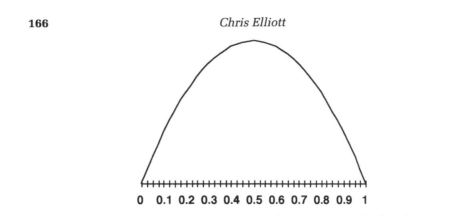

Fig. 10 Effectiveness of a train as a function of engine fraction.

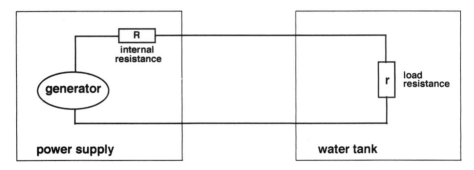

Fig. 11 An electrical heating circuit.

resistance of the power supply, R, even though half of the energy is then dissipated in the internal resistance.

A more subtle example can be found in nature. Photosynthesis is the process by which plants convert light energy into stored chemical energy. It is a seven-stage process and around 0.8 per cent of the incident energy ends up being stored. If each stage is 50 per cent efficient, the overall efficiency would be expected to be $(0.5)^7$, which is almost exactly 0.8 per cent. It looks as though evolution has, not surprisingly, resulted in optimum **effectiveness**, rather than high efficiency.

Measures of effectiveness

Unfortunately engineers are rarely allowed the luxury afforded to evolution. They are supposed to build one solution and get it right, not build many different solutions out of which the most effective survives. They need a measure which allows the potential effectiveness of each possible solution to be predicted *before* it is built. Any real problem has many qualities which have to be considered and we have to combine them to calculate a score.

Let's start with a non-engineering example—which university is best?

The Times Good University Guide takes ten separate qualities and estimates how well each university scores on each measure. It then calculates a combined score for each university by adding together each of the individual scores. This is a way of bringing together what could be called 'hard' qualities, like the number of First Class degrees awarded or the staff/student ratio, and 'soft' qualities, like the standard of student accommodation.

The result that emerges is not to everyone's liking. The Chairman of the Committee of Vice Chancellors and Principals said 'The compiler has taken single characteristics and combined them in a highly specific way that will be irrelevant to the needs of many students.' An individual student might have particular needs or interest—sporting facilities, wheelchair access, specialization in a subject—and the combination of scores will not reflect that student's requirements. The general measure of goodness may suit many students but not all. This is a reflection of the more general observation that there is no such thing as absolute quality—only fitness for purpose. The measure of goodness that is specific to the needs of a particular customer or user is usually called a 'figure of merit'.

In its simplest form a weight is attached to each quality to reflect its importance to the particular customer or user. The figure of merit of a candidate solution is found by multiplying the score achieved for that quality by the weight attributed by the user and summing over all qualities. *The Times* used a weight of 1 for each of the qualities in its calculation. More complicated figures of merit can be defined, including:

- hard thresholds: 'I won't consider any university that does not score above 75 on student accommodation'
- combinations of qualities: 'I want a university with a high number of postgraduate students and excellent research, so I'll multiply those two scores together'
- non-linear rules to prevent a single quality preempting the result: 'Any points over 90 for staff/student ratio will add half as much to the overall score as points below 90'.

The idea of a figure of merit is simple but its implications are far-reaching. The crucial point is that it makes the decision-making process into a one-dimensional problem. All the time that you have two separate measures of goodness, you have no rational basis for making a choice. Purists are uncomfortable with the idea of combining dissimilar qualities but there is no choice. Any choice has to weigh up the relative merits of the different qualities—a figure of merit is a way of making that process objective and visible.

Once a figure of merit has been defined, it becomes a design tool. The engineer can estimate the score of each of the candidate solutions before building them. More importantly, it becomes possible to make a rational trade-off between conflicting qualities. One example of a trade-off is between price, performance, and schedule. It is usually possible to satisfy any two of these but only at the expense of the third. So, an engineer could build the solution that does everything you want and could deliver it when you want it, but the cost will be excessive. If you insist on reducing the price, either the quality or delivery date will have to slip.

A more contentious trade-off involves safety, cost, and performance. Statements like 'safety is paramount' are meaningless. A completely safe train would never leave the station. Instead we have to use a figure of merit to trade between these three qualities. For example, the UK Department of Transport uses a rule that it is worth spending around £750 000 on road improvements to save one life.

Trade-off and decision making are at the heart of engineering design. There is no 'right' answer—the aim of the designer is to find the least bad compromise.

Systematic design

The principles of decision making that I have developed now allow us to challenge Sir Ove Arup's claim that possible solutions cannot be compared by quantitative methods. We can apply those principles by setting out to solve any design problem by answering three questions:

- what is the problem?
- what solutions are possible?
- which is the least bad?

What is the problem?

This might seem to be the easiest question to answer but often is the hardest. A useful guiding principle is 'the customer is rarely right'. Customers frequently pose their requirements in terms of prospective solutions and it is part of the designer's job to find the true underlying need. A famous example will illustrate this.

During the *Apollo* moon landing programme, it was realized that the astronauts would need to write notes and that a normal ball-point pen needs gravity to make the ink flow. NASA invested millions of dollars in developing a pressurized pen that would work in free-fall. Many years later, in an early period of *detente*, an *Apollo* capsule linked to a Russian

Soyuz capsule in orbit. The cosmonauts came drifting through the hatch carrying not only a sociable bottle of vodka but also a pencil.

The mistake was to specify that what was required was a ball-point pen that would work without gravity, rather than a writing implement that would work without gravity. The lesson is that the engineer can only define the correct figure of merit after a thorough analysis of the customer's real needs. In particular, he has to identify what qualities the customer requires and what weights to give them.

What solutions exist?

In many ways this is the easiest question, not because it is intrinsically easy but because it is what everyone recognizes as design. Research and development feed into the inventive and innovative process of devising candidate solutions. I only want to draw attention to one danger, which brings us back to the subject of camels.

We considered earlier the problem of optimization, which I illustrated as a hump—the performance of the solution varies as some parameter is changed and the aim of the designer is to get to the top of the hump. That is fine if one is engaged in dromedary engineering, where there is only one hump. However, many engineering problems are bactrian— they have two humps (see Fig. 12).

Engineers are generally very skilled at making incremental improvements to an existing design. They are less good at finding the correct starting point. Figure 12 might, for example, describe the evolution of personal computers—the first attempt was DOS and the successive improvements were Windows and Windows 95. Meanwhile, the Mac alternative was there all the time and, even in its first incarnation, was a better solution.

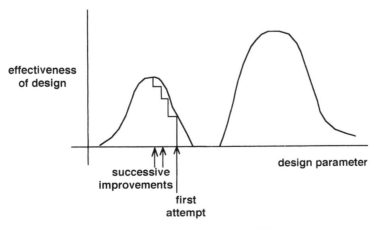

Fig. 12 A 'bactrian' design problem.

A current engineering problem is that of the electric car. Huge efforts are going into developing ever better batteries. The essential quality of a battery is that it converts chemical energy into electrical energy and is recharged by reversing the process—electricity is pushed back into the battery which is converted into stored chemical energy. The alternative is a fuel cell which also converts chemical energy to electrical energy but is recharged by adding new chemicals. The potential performance of the fuel cells corresponds to the hump on the right of Fig. 12 but many researchers are struggling on with small improvements to batteries, corresponding to the hump on the left.

It should now be even clearer how the story of NASA and the ball-point pen could have happened. If the problem is incorrectly defined, it will exclude the best solution, that on the right in Fig. 12, without it even being considered.

Which solution is least bad?

At this point the designer's job gets easy. Question 1 gave him a figure of merit, Question 2 gave him a range of candidate solutions, and all that is left is to calculate the effectiveness of each candidate and select the least bad.

Two points are worth noting. If one solution is obviously much worse than another, do not waste time trying to work out exactly how much worse—save the effort for the viable candidates that need more careful analysis. If two candidate solutions have similar scores, do not let the figure of merit determine the choice. It is generally a 'rough and ready' tool and not accurate enough for fine decisions. The precautionary principle, 'how could this solution fail?', is more important, so always think about the fall-back position if something goes wrong.

Detailed design

The process that I have been describing is the one that is needed before starting detailed design. Its aim is to find the general form of the least bad solution and leaves a different and in many ways much harder job requiring different talents and skills. However, it is a job for which there are many well-developed design tools, from computer-aided design and simulation through to modelling and field trials.

I hold those who can do that job in the greatest respect since I know that I could not do it well but, unless the system design is right, they will end up with a beautifully engineered wrong solution. The American writer Robert Benchley summed up this argument very well when he said that he thought that the most difficult part of building a bridge must be starting it: 'I might be able to finish it if someone would start it for me, but as for the first move, I would be left blushing furiously.'

Complex decisions

Systems in practice

The principles that help when taking simple decisions (the two sorts of mistake, the precautionary principle) and complicated decisions (what is the problem?, what solutions exist?, which is the least bad?) can be applied in the context of complex decisions about systems—a set of components which, when they are brought together, exhibit properties that were not present in the components alone. Now we have to consider that the decision that you take about one of the elements may have an impact on those emergent properties and thus depends on bigger issues than the element alone.

Let's start with a simple system. The components are a battery, a light bulb, a red wire with clips on its ends, and a similar black wire. If I clip one end of the red wire onto one terminal of the battery and the other end onto one of the tags of the light bulb, and clip the black wire onto the other battery terminal and the other tag of the light bulb, the bulb glows. The emergent property is light, which was not there in the components before they were brought together. I could replace the black wire with one that outwardly looked the same, but was in fact only the plastic sheathing without the vital copper core. Then no light would be produced, even though the system appears the same as before. This demonstrates that it is necessary to understand how the system works **as a system** in order to select the right components.

During the Second World War, the natives of Melanesia were fascinated by the behaviour of visiting troops. These strangers levelled a section of forest and painted some white lines on it, then machines landed from the sky and unloaded beer and cigaretees. After the war ended, the 'Cargo Cult' persisted in preparing airfields but, because they had no understanding of the system of which airfields were part, were unable to understand why no beer and cigarettes arrived.

It is not necessary to go so far from home to find another example of the 'Cargo Cult' mentality. The British Home Secretary recently announced that part of a policy on law and order would be the introduction of mandatory life sentences on a second conviction for rape. Wise judges immediately pointed out that there can be no sentence without a conviction and, if the only possible sentence is life, juries will be reluctant to return guilty verdicts unless the evidence is absolutely certain. The result will be that on average rapists will receive lighter, not more severe sentences. A further twist is that, since the sentence will be the same for rape and murder, a rapist might as well kill his victim to prevent her identifying him.

The judicial system has many elements of which statutory sentencing policy is only one. The Home Secretary had, like my using the black 'wire' or the Cargo Cultists building runways, latched on to one element of the problem without considering the system as a whole.

Similar examples abound. A very simple one occurred recently in environmental legislation. Toxic zinc waste used to be recycled and used in new products. It was then declared to be a dangerous substance so, instead of being recycled, must be dumped in landfill sites, thus increasing the environmental damage—the exact opposite of the intention of the legislation. Systems are like snakes—if you grab hold of the tail, the head twists round and bites you.

One more general example will serve as a useful introduction to the more detailed case studies that follow. This comes from farming. The infectious disease sheep scab has been largely eliminated by the compulsory dipping of sheep. All sheep were dipped and there was no opportunity for the disease to persist. Dipping was then made optional. Each farmer had to trade off the costs of dipping against the risk of infection (a simple decision by my definition). Even following the precautionary principle, most would correctly come to the conclusion that it was no longer worth dipping. The result was the reappearance of sheep scab because there was now a large number of undipped sheep in which it could thrive. The significance of this example is that each farmer's decision not to dip was correct, given the costs and risks involved. It is only when all of the farms are considered together that the problem arises.

Transport systems

The map in Fig. 13 shows a typical commuter choice. There are two roads from Suburbia to The City, each passing through a village. The road from A to B and the road from C to D are fast dual carriageways and the average journey time is 30 minutes. The roads from A to C and from B to D pass through the villages and are congested. This means that the time taken to drive them depends on the number of cars using them. Assume that the time taken is 2 minutes for every car per minute so, if 5 cars per minute are going through a village, the time taken is 10 minutes.

Now we will assume that 16 cars per minute travel from Suburbia to The City at rush hour. After a few days, the drivers will learn which route is quickest and the traffic flow will settle down to 8 cars per

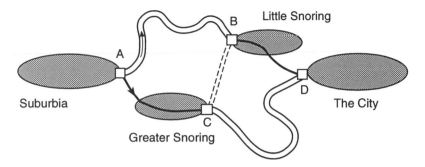

Fig. 13 A road network.

minute using each route. The journey time for each driver will be (2×8) $+ 30 = 46$ minutes. This is a state of equilibrium, where no driver can reduce his journey time by changing his action.

Now let's assume that we build the Snorings bypass from B to C, to reduce the traffic through the villages (dashed line in Fig. 13). The bypass is also a dual carriageway and uncongested, taking 4 minutes to drive. On the first day that it is open, let's assume that one driver per minute who reaches C takes the bypass. He has driven through Greater Snoring (16 minutes), along the bypass (4 minutes) and through Little Snoring (18 minutes since there are now 9 cars per minute through that village). His journey time is 38 minutes, significantly shorter than before. The others at C who go straight on to D still take 46 minutes but those who went from A to B now take 48 minutes because of the extra car's contribution to congestion in Little Snoring.

It is still worth taking the bypass from C to B and more cars will do it. Eventually, when equilibrium is reached, 13 cars per minute go through each village and each journey takes 56 minutes. The result of the bypass is a 60% *increase* in the traffic through each village and a 10 minute *increase* in everyone's journey time.

Is this a realistic scenario? It has been suggested that this might be a consequence of the Newbury bypass. This phenomenon was first predicted in the 1960s by Braess, a German mathematician, and some of his colleagues claim to have observed it in Stuttgart as early as 1968. You might also recognize the numbers. They appeared at the start of this Discourse in order to explain the surprising behaviour of the springs (shown in Fig. 1) when the string is cut. Cutting the string is analogous to closing the bypass and the equations are identical.

Incidentally, you will probably realize that I chose the numbers in these examples carefully. Once the rush hour traffic rate exceeds 30 cars per minute, the bypass *does* become effective. Similarly, a greater weight is hung on the springs, it would fall when the string was cut. The messages that I am trying present are that:

- you need to analyse a system carefully before making a change to be sure that your change will have the effect that you desire
- the free market in which every player takes the decision which is best for him locally (be he a driver or a sheep farmer) may give a result in which everyone loses.

Now let's look at a public transport system. Imagine a railway line with a rail network of six stations, A, B, C, D, E, and F, in that order along the line. I shall assume that 1000 people per day wish to travel between any two stations (so 1000 travel A to B, another 1000 travel A to C, yet another 1000 travel B to C, and so on) and that the fare that they will pay corresponds to £1 per stop for the return journey. I shall also assume that the cost of running each section of track is £3000 per day, except the section from C to D which costs £10 000 because of its difficult construction. The total income earned by the network is £35 000 per day and the total running cost is £22 000 per day—clearly a viable proposition.

Now what happens if we split up the network and make each section self-accounting? Section A–B and E–F have an income of £5000 per day (all the income from the 1000 passengers travelling from A to B, half the income from the 1000 passengers travelling from A to C, a third of the income from passengers travelling from A to D, etc.) and cost £3000, so they are still viable. Sections B–C and D–E have an income of £8000 per day and a cost of £3000—they are even better. Section C–D has the highest income, £9000 per day, but costs £10 000 so is uneconomic. If it has to be truly self-accounting, it must close. Now what happens to the other sections? Their passenger numbers all fall and the income is only £2000 per day each. Since each costs £3000 per day to run, they too must close. The previously viable network has collapsed because of incremental accounting decisions. Note that this has nothing to do with ownership. If the sections were each privately owned and operated, the owners of the outer sections could get together and subsidize the operator of C–D for their mutual benefit but that requires one to analyse the network as a system. Piecemeal analysis is no different from the black 'wire' or the Cargo Cultists' airfields.

In conclusion

I have tried to argue that there are simple decisions (which have two sorts of error and you can choose which you prefer), complicated decisions (where it is not even obvious what a good solution looks like), and complex decisions (where the perverse behaviour of systems means that your decisions can only be made in the light of many other decisions).

This is the world inhabited by the design engineer. I started with a quotation from an eminent structural engineer and tried to show that perhaps he was unduly pessimistic. I shall finish with a quotation from another eminent structural engineer, but this is one that I cannot fault:

> Engineering is the art of moulding materials we do not wholly understand into shapes we cannot precisely analyse, so as to withstand forces that we cannot really assess, in such a way that the community at large has no reason to suspect the extent of our ignorance.
>
> (Dr A.R. Dykes, Institution of Structural Engineers, 1976)

CHRIS ELLIOTT

Born 1952 and educated at Rendcomb College in Gloucestershire, Trinity Hall, Cambridge, and the Cavendish Laboratory, Cambridge where he was awarded a PhD for research into turbulent flow. He joined Smith System Engineering in 1976 and has worked in problems ranging from electronic warfare and automatic fingerprint recognition to the management and economics of space projects. In 1991 he was appointed a Royal Academy of Engineering Visiting Professor of the Principles of Engineering Design at the University of Bristol, an appointment generously sponsored by Charter plc. He conspires with three fellow professors drawn from civil engineering, aircraft and car design to sow seeds of scepticism, creativity, and, above all, excitement in the minds of engineering undergraduates. In 1996 he left Smith System Engineering to pursue a new career as a barrister, specialising in environmental law.

Safety's debt to Davy and Faraday

JAMES McQUAID

There are many direct connections between observations made by Sir Humphry Davy and Michael Faraday and concepts and devices of great importance to the safety of modern industry. Some of these developments now have connections beyond safety although they originated from the interest of Davy and Faraday in safety in coal mines. The only connection most people would make between Davy and safety is the miner's safety lamp, and Faraday's connection would probably be surmised as deriving only from his role as Davy's laboratory assistant. This paper will seek to bring out the importance of some of their separate observations to safety precautions and to environmental protection.

The involvement of Davy and Faraday in safety

Davy and Faraday each became involved in safety studies in different ways. Davy was approached on behalf of a society formed in Sunderland in October 1813 following the explosion at Felling colliery, Co. Durham, on 25 May 1812 in which 92 men and boys died. This was the largest death toll in a mine explosion up to that time. The Sunderland Society was a response by concerned individuals, including Members of Parliament, the nobility, and churchmen, to the need to press for a solution to the dreadful effects of explosions of firedamp, the flammable gas emitted when coal is mined. They took advice from John Buddle, the most eminent mining engineer of the day. Buddle's advice concluded that 'it is to scientific men only that we must look up for assistance in providing a cheap and effectual remedy'. This advice was followed up in August 1815 by the Rector of Bishopwearmouth, the Reverend Dr Robert Gray, who wrote to Davy and asked him to investigate solutions to the problem of firedamp ignitions by the naked candles used by miners as their source of light. He immediately assented and shortly afterwards paid a visit to Newcastle upon Tyne. Within two weeks of starting work, he had established the principle of a lamp which would provide a safe means of lighting. By the end of the same year, after many experiments to perfect the design, he had developed a practicable lamp. The lamp was first tested underground at Hebburn colliery on 9 January 1816, only five months after Davy had received Dr Gray's letter. He published his findings in a series of papers in which he included the observations he had made in the course of the collateral investigations to which his researches on the lamp had given rise. The development of the lamp was a true scientific process in which, in Davy's words, 'every step was furnished by experiment or induction, in which nothing can be said to be owing to accident'[1]. He used coal gas rather than firedamp in his experiments in order to subject his proposed solutions to a more severe test. He reported his observations extensively and noted that 'the results of these labours will, I trust, be useful to the cause of science, by proving that even the most apparently abstract philosophical truths may be connected with applications to the common wants and purposes of life'[1]. But he could hardly have guessed just how wide those applications would be. His work in that short period has proved to be remarkably influential and deserves to be better known other than in the name of a lamp that has now largely been superseded. He was generous in his acknowledgements to Mr Michael Faraday 'for much able assistance in the prosecution of my experiments'[1].

Faraday's direct connection with safety also came about as a result of a colliery explosion. The explosion was in 1844 at Haswell colliery, again in Co. Durham, and 95 miners died in the explosion. As a result of petitions to the Queen and the Prime Minister, Faraday was commissioned with the eminent geologist Sir Charles Lyell to investigate the explosion. They produced a lengthy report with perceptive observations on mining practices and recommendations to prevent a recurrence. The report became a matter of controversy with the mine owners who derided the recommendations as based on invalid assumptions due to the Commissioners not having practical experience of mining. One particular observation, that burning coal dust could propagate an explosion, continued to be the subject of fierce controversy until finally proved conclusively towards the end of the century. Lyell and Faraday's recommendations on other matters also came to be adopted at a much later date and one in particular can justifiably be said to underpin the safety of high production coal mines at the present day. As a bonus, it lessens the effect of underground mining operations on global warming—a connection that Faraday could not have imagined.

Coal mining in the days of Davy and Faraday

By the early nineteenth century, the mining of coal had progressed from extraction of the shallow seams near where they outcropped at the surface to deeper seams accessed by vertical shafts. Depths of workings of several hundred metres were becoming common. In the case of the Haswell colliery explosion, the seam in which the explosion occurred was at a depth of nearly 300 m.

A major source of danger in the mines came from firedamp, a flammable gas evolved from the coal. Firedamp, or marsh gas, is one of the products of the decomposition of vegetable matter. With shallow seams, the gas was very rarely encountered simply because over time it escaped to the surface through fissures in the strata. As a general rule, the deeper the mine, the more firedamp is released into the mine atmosphere as the coal is extracted. So firedamp became more prevalent and firedamp explosions began to be the order of the day as deeper seams were worked in the early years of the nineteenth century. We only know of the more calamitous explosions killing tens of miners at a time but smaller explosions killing only those in the immediate vicinity must have been a common occurrence.

As the firedamp hazard increased, so the need for better ventilation was appreciated. Systematic ways of laying out the way the coal seam was worked began to be practised[2]. Ventilation systems became very extensive; in some cases the air travelled up to 75 miles through the workings. The means for providing the ventilation were rudimentary, with efficient mechanical fans still some way into the future.

The earliest ventilation systems relied on the natural circulation induced by thermal effects. The temperature of strata varies with depth, typically being constant for the first 20m or so and thereafter increasing by 1°C for every 50 m, although these figures can vary substantially depending on the thermal properties of the strata. Some mines in the UK had strata temperatures over 30°C at the maximum operating depth of around 1000 m. The effect of the increased temperature underground was to heat the air and cause it to flow from one shaft to the other. The magnitude of this natural ventilation current varied through the year, being greatest in winter. Indeed, it was noticed early on that firedamp explosions were much less frequent in the winter. But, as the mines became deeper, there was a reversal of this annual pattern and explosions became more frequent in winter. The reason for this curious effect will be given later, but suffice to say that the explosions in the latter case were being propagated by coal dust rather than firedamp.

An alternative way of controlling the firedamp was obviously, though also dangerously, by firing it. It had long been known that firedamp was lighter than air and hence tended to accumulate in roof cavities and the higher parts of inclined workings. It became the practice, before each working day, for one of the miners to enter the workings with a candle on the end of a long pole. He progressively fired the gas that had accumulated overnight. He was clothed in wet sacking to give him some protection and his appearance, reminiscent of monks, caused him to be known as the *penitent* in France (Fig. 1).

The vagaries of natural ventilation with the weather led to the introduction of furnace ventilation. A permanently maintained furnace was installed to boost the circulation of air. The furnace was usually at the bottom of the upcast shaft up which the air left the mine, rather like a fire at the base of a chimney. Ventilation by furnace persisted for a long time and, as recently as the 1950s, small mines in the Forest of Dean were ventilated by a fire bucket suspended in the upcast shaft.

Davy was well aware of the hazard of firedamp. He observed that 'coal, when the pressure of the superincumbent material is removed,

Fig. 1 The *penitent* firing firedamp.

affords inflammable air which is disengaged not only when the coal is broken but is likewise permanently evolved, often in enormous quantities, from fissures in the strata'[1]. He further noted that 'Sir James Lowther had observed early in the last century that the firedamp was not inflammable by sparks from flint or steel; and a person in his employment had invented a mill for giving light by the collision of flint and steel and this was the only instrument except common candles employed in the dangerous parts of the British collieries'[1]. The Spedding mill, as it was known after its inventor, became the standard method of illumination in the presence of firedamp (Fig. 2). But explosions due to ignition of firedamp by the Spedding mill did in fact occur and this was a reason for the approach to Davy for development of a safe means of lighting.

Another suspected hazard was coal dust. It was reported that survivors of an explosion at Wallsend colliery in 1803 were burnt by a shower of red-hot sparks even though they were remote from the origin of the explosion. But it was not until Lyell and Faraday investigated the

Fig. 2 The Spedding mill as a source of light.

Haswell explosion that an explanation was given of the manner in which coal dust might contribute directly to the violence of an explosion.

The social conditions of the times were not conducive to action to improve the lot of the miner. There was no legislation to regulate the conditions in the mines. Attempts to arouse public concern lay in the hands of committees, such as the Sunderland Society and, later, the South Shields Society. A Select Committee of the House of Commons carried out an investigation in 1835 and concluded that it was of the greatest importance to provide ventilation sufficient to dilute the firedamp so that it was no longer explosive. They relegated the safety lamp to second place, looking upon it as an aid to the miner. There was no government inspection until the 1840s and then only to regulate the employment of women and children, under an Act passed in 1843. The inspector appointed under that Act advocated inspection of the mode of ventilation in each colliery by a properly qualified Government Officer but without powers of interference. He opposed compulsion on the grounds that this would transfer responsibility for safety from the owners to the Government. There was no organized scientific research into mine safety and the standard practice, after a major explosion, was to appoint scientific men to carry out an investigation, as with Lyell and Faraday after the Haswell explosion. Others who were called upon in a similar way were Sir Lyon Playfair and Sir Henry de la Beche.

Davy's investigation of the safety lamp

Davy expressed his remit in the following terms: 'The great object, one ardently desired rather than confidently expected, was to find a light, which at the same time that it enabled the miner to work with security in explosive atmospheres, should likewise consume the firedamp'[1]. Davy tackled the problem from first principles. He carried out a chemical examination of the firedamp using samples obtained from mines. He confirmed what was already known, that the firedamp was largely composed of methane. He then made numerous experiments on the circumstances under which it explodes and the degree of its flammability. He found that mixtures of air with too much or too little firedamp would not burn. If there was too little firedamp, the flame of a taper was merely enlarged in the mixture, an effect which was still perceived in thirty parts of air to one of gas. Davy's studies led to the establishment of what are now known as the flammability limits of a gas, which define the range of concentration within which self-sustained combustion is possible. The concept of flammability limits is central to the control of the hazard of explosion of flammable gases as well, of course, as the design of all types of gas combustion equipment.

Davy then observed that firedamp was much less combustible than other flammable gases. By this he meant that it was not easy to ignite. Specifically, he found that it was not exploded or fired by red hot charcoal or red hot iron. This property of difficulty of ignition forms the basis of intrinsically safe electrical apparatus whereby it is possible to design electrical apparatus so that the sparking produced at any break of the circuit is intrinsically unlikely to ignite a specified flammable gas, with different gases being easier to ignite than others. For example, hydrogen is much easier to ignire than firedamp. Coal gas, or mains gas as it used to be known, produced from gasification of coal, is easier to ignite than methane because of its hydrogen content. This observation will be familiar to those who remember the conversion of domestic gas supplies from mains gas to North Sea or natural gas (i.e. largely methane) in the 1970s.

Davy found that the heat produced by firedamp in combustion was much less than that of most other flammable gases. It was then a natural development for him to investigate how flames were transmitted in various circumstances. He first experimented with flames transmitted through tubes. He found that, if the tube was small enough and long enough, a flame would not pass through the tube. Furthermore, metallic tubes prevented flame transmission better than glass tubes. Davy was now close to perfecting the principle of the safety lamp. He summed it up thus: 'In reasoning upon these various phenomena, it occurred to me, as a considerable heat was required for the inflammation of the firedamp and as it produced in burning comparatively a small degree of heat, that the effect of the surfaces of small tubes in preventing its explosion depended upon their cooling powers; upon their lowering the tempera- ture of the exploding mixture so much that it was no longer sufficient for its continuous inflammation'[1]. This idea led to an immediate result— the possibility of constructing a lamp in which the cooling powers of the apertures through which the firedamp-contaminated air entered or the burnt gas made its exit should prevent the transmission of flame. Although his practical conclusion was valid, Davy was not correct in saying that the low heat of combustion was the significant parameter. Rather, the reason for the ease with which a firedamp flame is quenched by cooling surfaces derives from the low chemical reactivity of firedamp and hence the low speed of transmission of flame through the gas. This means that the rate of heat loss to the cooling surfaces can easily be made to exceed the rate of heat generation by the chemical reaction and, once this happens, the flame is extinguished. Davy's observations on flame quenching provide the basis for applications, often involving other flammable gases besides firedamp, where a flame can be extinguished by deliberate design of passageways along which the flame must travel. This is the principle of flameproof electrical apparatus and flame arresters that now find wide and essential application anywhere where a flammable gas might be present, for example, in oil refineries, offshore platforms, and the petrol pumps in every filling station.

Following his establishment of the practicality of flame quenching, Davy then experimented with many systems of tubes but concluded that small tubes or apertures were not effective. In rejecting these systems, he stated 'and at last I arrived at the conclusion that a metallic tissue, however thin and fine, of which the apertures filled more space than the cooling surface, so as to be permeable to air and light, offered a perfect barrier to explosion, from the force being divided between, and the heat communicated to, an immense number of surfaces'[1]. His solution was to surround the light entirely by wire gauze capable of feeding the flame with air and transmitting light. In plunging such a lamp into an explosive mixture, he saw the gauze cylinder become quietly and gradually filled with flame, the upper part soon appearing red hot, yet no explosion was produced.

At this stage, Davy reasoned that the samples of firedamp he used might not expose the lamp to the most severe test. He believed that firedamp might, in some circumstances, contain small quantities of olefins and these would increase the heat of combustion. He thereafter resolved to use coal gas, which also contains olefins and hydrogen, as the test gas. He then found the type of gauze which would make the lamp safe under all circumstances in atmospheres containing coal gas so that there would be a factor of safety, as we would now call it, when the lamp was used in firedamp. This was the lamp that was tested underground in January 1816 and immediately adopted.

The discovery of catalytic oxidation

Davy did not stop there and rest on his laurels. He realized that the flammability limits depend on temperature and so he proceeded to study the effect of preheating of the firedamp and air due to the heating up of the gauze. As a result, he undertook 'experiments on the nature of flame and the modifications of combustion [that] led me in January 1817 to an important practical addition, founded on an entirely new principle'[1]. He began by referring to the phenomenon of slow combustion without flame. He remarked that the temperature of flame is 'infinitely higher than that required for the ignition of solid bodies'[3]. In this, he uses 'ignition' to mean simply the heating of the solid bodies so that they emit visible light. It appeared to him probable that 'when the increase of temperature was not sufficient to render the gaseous matters themselves luminous, yet still it might be adequate to ignite solid matters exposed to them'[3]. What he was trying to do was to see whether the heat produced by slow combustion without flaming would cause solid matter to glow. He had devised several experiments on this subject when he was accidentally led to the discovery of a new and curious series of phenomena. What he then described was his discovery of catalytic oxidation.

In his experiments Davy introduced a small wire-gauze safety-lamp with some fine platinum wire fixed above the flame into a mixture at the upper flammability limit. When the flame took hold in the wire-gauze cylinder, he introduced more coal gas expecting that the heating of the mixed gas passing through the wire gauze would prevent the excess from extinguishing the flame. In other words, he expected that the upper flammability limit would be increased by the increased temperature of the mixed gas. He noticed that the flame continued for only two or three seconds after the coal gas was introduced. When it was extinguished, a curious phenomenon occurred: the platinum wire continued to glow for many minutes. When the experiment was repeated in a dark room, it was evident that there was no flame in the cylinder. It was immediately obvious that the oxygen and gas in contact with the hot platinum wire were reacting without flame and producing enough heat to keep the wire glowing. Davy proved the truth of this conclusion by making a similar mixture, heating a fine platinum wire, and introducing it into the mixture. It immediately glowed nearly to whiteness, and continued glowing for a long while. When it ceased to glow, Davy found that the flammable gas had been completely consumed. He further observed that a temperature much below that needed to make the wire glow would also produce the phenomenon. If the wire was repeatedly taken out and allowed to cool until it ceased to be visibly red, and was then reintroduced, it instantly became red hot.

Davy repeated the experiments with other flammable gases and vapours. He then tried to produce the phenomenon with various metals but succeeded only with platinum and palladium. With copper, silver, iron, zinc, and gold, the effect was not produced. Davy surmised that the explanation lay in the low thermal conductivity and small heat capacity of platinum and palladium relative to other metals. Of course, he was wrong in that initial explanation of why it happened with platinum and palladium but not other metals. Catalytic oxidation is a much more complicated phenomenon. Davy remarked that many theoretical views would arise for connecting his observed facts and hints for new researches would also arise which he hoped to be able to pursue.

Davy described the practical application of his discovery thus:

> By hanging some coils of fine wire of platinum above the wick of his lamp in the wire gauze cylinder, the miner, there is every reason to believe, will be supplied with light in mixtures of firedamp no longer explosive; and should his flame be extinguished by the quantity of firedamp, the glow of the metal will continue to guide him and by placing the lamp in different parts of the gallery, the relative brightness of the wire will show the state of the atmosphere in those parts[3].

I have dwelt on this particular aspect of Davy's work on the safety lamp at length both because of the insight it gives into an important scientific discovery and because it has a direct connection with a much later development of equal importance to the safety lamp in the effort to control the firedamp hazard. That development took place at the Safety in Mines Research Establishment in the 1950s and resulted in the pellistor, a device relying on the catalytic oxidation of the firedamp by palladium. The pellistor forms the basis of the modern instrument for measuring firedamp concentrations in mines and in many applications in industries where flammable gases are a hazard, such as oil refineries and offshore platforms. A pellistor is a miniature transducer in which a flammable gas is oxidized on a small ceramic bead coated with a palladium catalyst and heated by a coil of platinum wire embedded in it (Fig. 3). The heat of oxidation raises the temperature of the bead and the rise of temperature results in a change in the electrical resistance of the platinum coil, though this platinum coil itself plays no part in the oxidation. The change of resistance is registered as a change of voltage across the coil and suitable calibration converts the change of voltage to concentra-

PLATINUM   **ALUMINA COATED**
WIRE COIL   **WITH CATALYST**

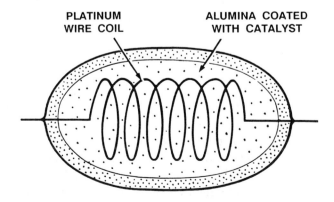

Fig. 3 The construction of the pellistor.

tion of the flammable gas. Some designs of pellistor use mixtures of palladium and platinum as the catalyst for different applications.

Prior to the development of the pellistor, attempts had been made to utilize catalytic oxidation to provide an indication of the presence of firedamp. In the late nineteenth century, two separate strands of development took place. Firedamp indicators developed in France used electrically heated platinum or palladium coils to burn the firedamp in a fixed volume of sampled mine air. In the version developed by Castel (4), the change of volume of the sample at constant pressure was calibrated against the concentration of firedamp, whilst that developed by Le Chatelier (5) used the change of pressure at constant volume. An early form of electric miner's lamp developed in Belgium by Sussman incorporated a firedamp detector (6). It consisted of a perforated metal cylinder housing a glass tube containing mercury. Finely divided platinum was placed on the outside of the glass tube. In the presence of firedamp, the heat of oxidation caused the mercury to expand, closing two electrical contacts within the tube and lighting a small red warning lamp. Later attempts to use a platinum coil directly as a combined resistance thermometer and catalyst were not generally successful due to lack of robustness resulting from the high temperature (around 1000 °C) at which the platinum coil had to operate to give a measurable response. The problem was successfully overcome by the construction of the pellistor, in which palladium replaced platinum as the catalyst and the catalytic oxidation component was separated from that providing the heating and measuring the temperature change. The pellistor provided a robust and reliable transducer with a higher surface area for oxidation than a coil and a lower operating temperature. It has since become the basis for standard handheld instruments for the supervisory staff in mines as well as for extensive automatic detection systems monitored from the surface control room. It finally brought to practical fruition in the aid of the miner the application of the 'entirely new principle' first observed by Davy in his researches on the safety lamp.

The safety lamp was, and occasionally still is, used as a means for detecting firedamp, long after Davy's original invention of it as a safe source of light had been overtaken by the development of safe electric lighting. As remarked earlier, Davy had observed that in weak non-explosive mixtures of firedamp and air, the flame of a taper was enlarged in the mixture. As is illustrated in Fig. 4 the flame of the lamp assumes a different appearance depending on the concentration. Experienced miners are able to use the safety lamp as an instrument for measuring the concentration of non-explosive mixtures. It is of course necessary to be able to detect these low concentrations as a warning of impending danger.

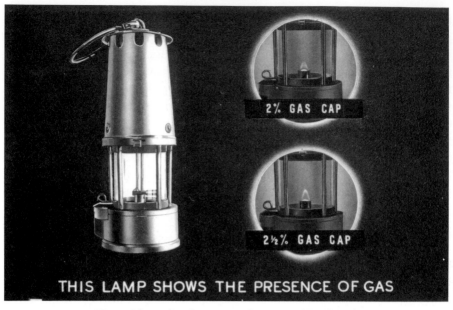

THIS LAMP SHOWS THE PRESENCE OF GAS

Fig. 4 The safety lamp as a detector of firedamp.

Thus there is a linear progression from Davy's original work on a safe lamp for lighting to its use as the miner's firedamp detector and then to its replacement by the modern application of Davy's discovery of catalytic oxidation.

Aside from safety, catalytic oxidation has many other applications, perhaps the best known being the catalytic converters that are now required to be fitted to motor car exhaust systems to reduce atmospheric pollution. Another contribution to reduced vehicle emissions comes from the use of additives to increase the oxygen content of unleaded petrol, thereby promoting more complete combustion. The production of these additives relies on the use of a platinum catalyst in the production process. The use of platinum catalysts is now widespread in industry, with applications ranging from the production of nitric acid needed for fertilizers, explosives, and synthetic fibres to pharmaceuticals essential to modern living such as antidepressants, oral contraceptives, and paracetomol.

Flameproof electrical equipment

Returning to Davy's other observations, I will first track the connection of two of them, both relying on the principle of the safety lamp i.e. the quenching effect of heat sinks on flames. Electrical equipment for power applications such as motors and switchgear is widely used in places where a flammable atmosphere may be present. Such equipment inevitably produces arcing at 'make and break' contacts. The thermal energy in these arcs is quite sufficient to ignite firedamp and, from Davy's own observations, other flammable gases will be ignited even more readily than firedamp. There are various ways of designing this kind of electrical equipment so as to avoid ignition of the external atmosphere. The commonest method is to follow the principle of the safety lamp by allowing the flammable atmosphere to enter the equipment but not allowing a flame to reach the external atmosphere; that is, the enclosure is made 'flameproof'. The way to do this was provided by Davy when he said:

> metallic doors or joinings in lamps may be easily made safe by causing them to project upon and fit closely to parallel metallic surfaces. Longitudinal air canals of metal may, I find, be employed and a few pieces of tin-plate soldered together with wires to regulate the diameter [presumably meaning the depth] of the canal, answer the purpose of the feeder or safe chimney[7].

Davy's description is still as good a definition of flameproof design as one can find. It means that the gap between the flanges of the enclosure of the electrical equipment is designed to ensure that a flame expanding through the gap will be extinguished by the cooling effect of the metal flanges and will not ignite the external atmosphere. An important extension of understanding of the phenomenon[8] was the realization that the jet of combustion products emerging from the gap, even though there was no flame, could ignite the external atmosphere. Taking this into account, the actual dimensions of 'safe' flange gaps have been established for different flammable gases and are laid down in standard specifications.

Provided the gaps are maintained as originally designed (for example, by tightening a requisite number of bolts to achieve the same effect as Davy's soldered wires), flameproof equipment is extremely reliable in achieving its intended function, although explosions have occurred as a result of the use of incorrectly maintained equipment. The explosion at Cambrian colliery in 1965, the last large-scale explosion in the UK mining industry (killing 35 miners), was attributed to the ignition of a firedamp accumulation by defective electrical equipment. However, as far as is known, no firing of an external atmosphere has ever been observed with a properly designed and maintained flameproof enclosure. The development took place over a long period, being first proposed around 1890, though the term 'flameproof' was not defined in a standard until 1924 when testing and certification of equipment began at the University of Sheffield (6).

Figure 5 shows a flameproof enclosure in which, paradoxically, flame transmission is taking place. There is a particular reason for this. The flammable gas did not enter from the external atmosphere but was generated internally by the decomposition of plastic components within the enclosure. The combustion properties of the gaseous products of decomposition were different from those of the gases for which the enclosure was designed to be flameproof, resulting in inadequate cooling by the flanges. Figure 6 is more typical and shows a high-powered coal-cutting machine powered by a flameproof electric motor on a modern coal face.

Flame arresters

The second widespread application of Davy's observations on quenching of flames is to devices known as flame arresters. In some circumstances, a flammable gas may be ever present in a system. A common example is the vapour venting system of a storage tank for flammable liquids such as petrol. There is increasing concern that the common practice of venting hydrocarbon vapours to the atmosphere is contributing substantially to

Fig. 5 A 'flameproof' enclosure transmitting flame.

Fig. 6 Coal-cutting machine powered by flameproof motor.

environmental harm both as a cause of urban smog and as a contributor to global warming—the greenhouse effect. A common application is at ship tanker loading facilities at oil refineries. As a ship's tank is loaded with liquid hydrocarbon, the vapour above the liquid in the tank must be vented. Until quite recently, it was standard practice to vent the vapour direct to the atmosphere. Nowadays, it is a requirement that the vent should be connected back to the refinery and the vented vapours processed or burned safely in a tall flare stack. The problem with this is that at times there is a long, large diameter pipe filled with flammable vapour. If this vapour was to be ignited for any reason, the results would be catastrophic. Since ignition cannot be ruled out as a possibility, it is necessary for safety that means should be provided to quench or arrest any flame transmitted along the pipe. The device for doing so is called a flame arrester. This relies on the cooling powers of metal elements placed in the pipe in the form of laminations, crimped ribbon, or compressed knitted wire mesh. The mass and geometry of the elements have to be designed to provide a small resistance to the flow of the vapour in normal operation but to be effective in quenching a flame under all circumstances.

A particular danger with flame arresters is the possibility of flame transmission due to sustained burning at the arrester, resulting in excessive heating of the metal elements and hence the loss of their cooling power. This danger was apprehended by Davy in his safety-lamp experiments. He noted:

> By heating strongly gases that burn with difficulty, their continued inflammation becomes easy, in consequence of increments of heat occasioned by combustion of small quantities, which under any other circumstances would not produce continued combustion. Hence if mixtures of firedamp are burnt from systems of tubes or canals or metallic plates, which have small radiating and cooling surfaces: though these systems are safe at first, they become dangerous as they are heated[1].

Davy described precisely the difficulty faced by the designer of a flame arrester. The avoidance of the problem in the case of the safety lamp is assured by the use of metal gauze with its high radiative properties. Indeed, Davy observed that 'even a red hot gauze, in sufficient quantity and of the proper degree of fineness, will abstract sufficient heat from the flame of firedamp to extinguish it'[1]. In the case of flame arresters, the actual arrester must be rigorously tested under conditions of sustained burning to ensure satisfactory performance; large test rigs are needed for this purpose.

Intrinsically safe electrical equipment

At various points in his description of the safety-lamp development, Davy referred to the difficulty in igniting firedamp. He noted the importance of ascertaining the 'degree of heat' required to explode the firedamp mixed with its proper proportion of air[7]. He found that a 'common' electrical spark, whose source he did not specify, would not explode five parts of air and one of firedamp though it exploded six parts of air and one of firedamp. 'Very strong' sparks from the discharge of a Leyden jar seemed to have the same power of exploding different mixtures of the gas as the flame of the taper. Davy was also aware that sparks from the Spedding mill were known to be capable of igniting firedamp, though the chance of this happening must have been small in view of the widespread use of the Spedding mill.

Davy's main interest was in the ease with which heated wires would ignite a gas. He found that:

an iron wire of one-fortieth of an inch heated cherry red will
not inflame olefiant gas but will inflame hydrogen. And a wire
of one-eight of an inch, heated to the same degree, will inflame
olefiant gas, but a wire of only one-five hundredth must be
heated to whiteness to inflame hydrogen. Wire of one-fortieth
of an inch heated even to whiteness will not inflame firedamp[9].

He concluded that these circumstances explained why a mesh of wire so
much finer was required to prevent the transmission of the explosion of
hydrogen and why so coarse a texture and wire was sufficient to prevent
the explosion of firedamp, fortunately the least ignitable of the known
flammable gases.

The lesson to be drawn from these disparate observations is that a
better understanding of the difficulty of ignition may be turned to
advantage. If the available energy for ignition, or 'degree of heat' as Davy
termed it, from known sources such as electrical sparks or hot particles or
surfaces can be assured, it should be possible by suitable design to keep
the energy available below that required to ignite the gas. Unfortunately,
the development of the necessary understanding to enable that design
concept to be applied to an acceptable level of safety required the stimulus of the disastrous explosion at the Universal colliery, Senghenydd,
Glamorgan in 1913 in which 439 miners were killed. This was the
largest ever death toll in a mining accident in Great Britain. The lessons
of that experience have been well learned and it is instructive to follow
them through, elaborating on Davy's original observations on the
ignitability of gases. The Senghenydd explosion involved the ignition of
firedamp by electrical sparks from low voltage equipment; the following
will concentrate on that particular ignition source and the precautions
that are now applied.

An electrical spark raises the temperature in a small volume surrounding it, initiating the combustion reaction[10]. Whether the reaction
is sustained depends on the rate at which the heat of reaction is lost to
the unburnt gas and to any adjacent heat-absorbing masses such as the
electrodes, as well as the time over which the temperature rise due to the
spark exceeds the normal flame temperature. The flame needs to grow to
a certain critical size in this time. If that does not happen, the rate of heat
loss will be greater than the rate of heat generation by burning. The
temperature will therefore fall, the combustion reaction will subside,
and only a small volume of the gas will be burnt. A minimum ignition
energy provided by the spark is required in order that the flame attains
the critical size and thereafter the combustion spreads throughout the
flammable gas. Thus weak sparks, of the type Davy referred to as

'common', are not suitable; instead, energetic or 'fat' sparks, such as those from Davy's Leyden jar, are required. This is very familiar experience with petrol-engined motor cars. Leakage of the electrical energy in the ignition system caused by poor insulation (for example from dampness) will make the engine difficult to start.

All of this was implicitly recognized in the adoption of unprotected low voltage signalling systems in mines early this century. These signalling systems were very simple, consisting of a trembler bell powered by a battery, as used in the common doorbell. The 'switch' of the bell was provided by the miner short-circuiting across bare wires which were strung along the wooden props of the haulage roads of the mines. He did this either by bringing the wires together or by bridging them with any strip of metal readily to hand. The act of short-circuiting produced a spark at the bell and at the place where the short was made or broken. The bell signal was used to alert another miner at a distant place for some purpose, usually to start up or shut down the haulage system. Such signalling systems were exempt from the provisions of the regulations governing the use of electricity in mines introduced in 1905.

A signalling system in use at the origin of the explosion at Senghenydd came under suspicion as the source of ignition. The evidence was inconclusive, although two explosions the year before at Bedwas and Caepontbren collieries in South Wales had definitely been attributed to ignition by a similar signalling system. The belief at the time was that ignition of firedamp could not occur if the voltage of the system did not exceed 15 volts. However, the investigations set in train after the Senghenydd explosion established that the inductance of the circuit was important as was the current, even more so than the voltage. The investigations led to recommendations on the design of bare-wire signalling systems and to the design of an apparatus which would test whether a circuit was capable of generating sufficiently energetic sparks under any conditions either in normal operation or under fault.

The use of electrical apparatus not only in coal mines but in all places where flammable atmospheres may be present requires that the design of the apparatus is such as to eliminate the risk of explosion due to sparking in the electrical circuits. The method of flameproof protection, already described, allows ignition by sparks to take place but prevents transmission of flame to the external atmosphere. A second method, based on the work initiated following the Senghenydd explosion, is to make the apparatus and related electrical circuits 'intrinsically safe'. This is in general limited in applications to apparatus in which the output or consumption of energy is small, as in the case of signalling systems. The term 'intrinsically safe' was adopted and defined in a

British Standard[11] in 1945. It denotes that apparatus is 'so constructed that when connected and used under the prescribed conditions any sparking that may occur in normal working, either in the apparatus or in the circuit associated therewith, is incapable of causing an explosion of the prescribed flammable gas or vapour'.

The ignition potential of sparks, or their 'incendivity', depends on the electrical characteristics such as voltage, inductance, current, and rate of current decay following a break, on the thermal characteristics of the physical surroundings at the spark, and on the flammability properties of the gas. It is still not possible to specify the relationship between incendivity and all these variables. The electrical design of intrinsically safe apparatus takes account of known factors that influence the total energy and the rate of energy input to sparks when a break occurs in a circuit. For example, if there is an inductive winding, it is necessary to divert or absorb the energy released when the circuit is broken so that the energy is not expended in the spark.

For assurance of safety, any intrinsically safe apparatus must be tested in a way that exposes the sparks generated at a break in the circuit to an atmosphere of the flammable gas in which the apparatus is intended to operate. However, a single test does not suffice. This is because there is an element of chance in whether ignition actually occurs, since the parameters that determine the outcome cannot be precisely controlled— an example of the limited predictability of what otherwise appears to be a deterministic outcome, as is now known to be true even, for example, in the case of systems that are governed by Newton's laws of motion[12]. In practice, it is customary to require at least 100 tests without ignition occurring before sufficient assurance of safety can be assumed. The term 'intrinsically safe' therefore equates to a very high assurance of safety rather than, as might be implied, to complete physical impossibility of unsafe operation. The equipment for carrying out these tests, known as the 'break flash apparatus', has been the subject of substantial development and refinement from the design developed following the Senghenydd explosion[13,14]. The design and testing of instruments and data transmission systems that are now widely used in installations where flammable atmospheres are likely to be present rely on understanding of the flammability and ignitability characteristics of gases whose foundations were so well laid by Davy's researches on the safety lamp.

The technology has recently been extended to the hazard associated with the use of optical sensing in flammable atmospheres. The fracture of an optical fibre carrying a laser beam will result in the heating of any dust particles or fibres on which the beam is incident. Ignition of a flammable atmosphere by such heated particles or fibres may occur, depending on the properties of the incident radiation and of the particles or fibres, as well as the flammability characteristics of the ambient atmosphere. Research has resulted in the development of a predictive strategy[15] for the avoidance of ignition so that optical systems can be specified so as to be intrinsically safe in an analogous way to electrical circuits.

The Haswell colliery explosion

On 28 September 1844 there was an explosion at Haswell colliery, about seven miles east of Durham. This was the event that resulted in Michael Faraday's direct involvement in safety. In conjunction with the geologist Sir Charles Lyell, he wrote a report[16] to the Home Secretary, Sir James Graham; it was dated 21 October 1844. The report has two themes—the geology of the mine and methods of preventing explosions. It is fair to assume that Faraday was responsible for the latter aspects. The Haswell colliery was in what is known as the concealed coalfield, overlain by magnesian limestone. It was a modern colliery, opened in 1831 and with a well-developed system of ventilation provided by two furnaces. Lyell and Faraday stated that Haswell 'contains as small a proportion of firedamp, and is as well ventilated as any in that part of England, surpassing in these respects most of the mines'. The Hutton seam in which the explosion occurred was at a depth of 300 m and was inclined at 1 in 24. The workings were very extensive, as shown in the plan in Fig. 7. The method of mining was that known as pillar and stall, as distinct from the longwall system nowadays practised. Candles were generally used as the source of light, with safety lamps in the more dangerous places.

With the pillar and stall system, roadways or stalls about 5 m wide were first mined, leaving pillars of coal about 16 m wide between them to support the roof. The pillars were later removed and the roof allowed to collapse. This resulted in the formation of large areas of collapsed ground known as the waste or goaf. The largest of the goaves at Haswell extended to about 5 hectares at the time of the explosion. Lyell and Faraday paid particular attention in their report to the effect of the goaves on safety in view of the fact that firedamp in large quantities accumulated in them. They observed in particular that the firedamp, being lighter than air, would fill up the higher parts of the goaf.

Lyell and Faraday argued that the extraction of a pillar of coal resulting in a fall of the roof would open a passage from the accumulation of firedamp into the mine airways. They made some calculations of the

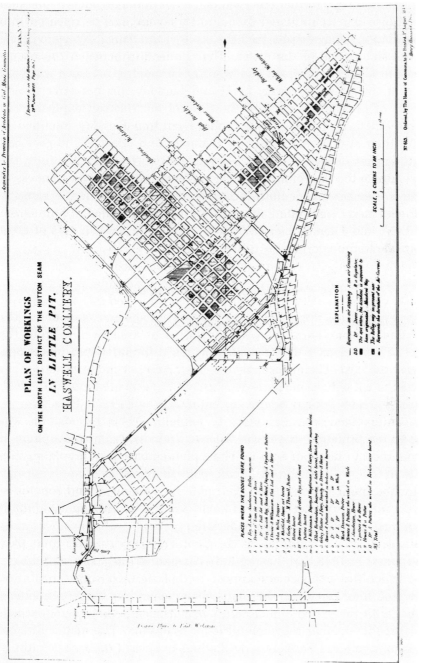

Fig. 7 The layout of the Hutton seam at the Haswell colliery.

volume of explosive mixture that would be liberated in foreseeable circumstances. They further observed that a fall in the barometric pressure would cause the accumulated firedamp to expand and be released into the ventilation. They emphasized the scale of the danger by again carrying out calculations of the effect. The connection between the occurrence of firedamp explosions and a fall of barometric pressure was well known at the time. However, Lyell and Faraday did not refer to any evidence of pressure changes at the time of the Haswell explosion. In carrying out their various calculations, Lyell and Faraday intended to support their strong belief that the character and effect of the goaf had been the cause of the explosion. They made two recommendations, the first to reduce the danger and the second to monitor the effectiveness of measures they proposed. Both recommendations subsequently came to be adopted many years later; although these were in different forms from what Lyell and Faraday had envisaged, they were nevertheless of great importance to the safety of the mines.

The principle of firedamp drainage

In order to reduce the danger, Lyell and Faraday first considered the possibility of ventilating the goaf by means of a shaft sunk from the surface. They rejected this, recognizing the difficulties and objections, not least the cost. Their report then states: 'Another mode of action has occurred to us, which, the more we think of it, seems the more practical, and offers greater hope of service to humanity, and which, therefore, we shall venture somewhat minutely to explain'. Their proposition was based on the principle that it was safer to drain away the atmosphere of the goaf than to blow air into it. Their plan consisted of laying an iron pipe from the goaf to the upcast shaft, with some means of applying suction to the pipe. They discussed their plan in some detail and concluded 'with some degree of confidence in its principles, we venture to submit it to practical men for their consideration'. Their confidence was misplaced. A Special Committee of the Coal Trade scrutinized their plan and reported in 1845 that 'having duly considered and explained the extreme difficulties, expense and almost, in their opinion, impracticability, of carrying into execution the plan recommended by those gentlemen, together with the extreme uncertainty of its success, they regret exceedingly that they cannot recommend it for adoption'. The matter did not end there; Lyell and Faraday were invited to submit their observations; it then transpired that they had not been consulted by the Coal Trade Committee and furthermore that Faraday had undertaken certain trials which he had described to the members of the Royal Institution at one of their Friday evening discourses. Lyell and Faraday would have sent the letter to the Coal Trade Committee 'had we been aware of the existence of the Committee, and it might, we venture to believe, have had some influence on their Report'. They responded to the various criticisms of their plan and, in tones of extreme politeness, declared,

> In making these remarks, we will wish them only to be viewed as suggestions made to practical men for their judgement and trial. It is a truth well known to every experimental investigator of nature that the success or failure of such a proposition as that made in our Report, often depends upon the animus with which it is carried into practice. Our first announcement was, if literally followed, very probably impracticable; but it was more the expression of a principle than a practice.

Lyell and Faraday's recommendation was the first official recognition that consideration should be given to a system of firedamp drainage. However, the principle was not entirely new and isolated examples of firedamp drainage from goaf areas were known, though evidently not to Lyell and Faraday or, even more surprisingly, to the Coal Trade Committee. For example, a Select Committee of the House of Commons in 1835 was informed of several instances of firedamp being piped from the goaves to the surface[2].

Lyell and Faraday's strong advocacy of firedamp drainage as being necessary to give sufficient assurance of safety lay fallow until almost one hundred years later. They had realized that the accumulation of firedamp in the goaf owed its origin not just to the firedamp released in the coal seam being mined but also to firedamp migrating from the disturbed strata above and below the seam. Whereas Lyell and Faraday proposed a drainage system to capture the firedamp when it accumulated in the goaf, the ultimate development has taken the form of boreholes drilled from the roadways into the strata above and below the seam in order to capture the firedamp before it enters the goaf and the mine roadways. The boreholes may be over 100 m long and are linked to a common pipe which is connected to a suction system at the surface of the mine. The first such system was installed in the Ruhr in 1943 and has since been widely adopted. The systems now in use capture around half of the total firedamp that would otherwise be released into the ventilation air of the mine. This has obvious benefits both to safety and to the economics of mining, allowing high production rates of coal without excessive ventilation costs. Furthermore, the drained firedamp is often used as a source of fuel for steam or gas turbines or reciprocating engines. Harworth colliery in Nottinghamshire, for example, generates 20 MW of power in this way[20].

An interesting aspect of the application of firedamp drainage is that the combustion of the firedamp significantly reduces the contribution of coal mining to global warming. Firedamp, being mainly methane, has a global-warming potential over a 100 year timespan which is 24.5 times that of carbon dioxide. Burning of the firedamp produces carbon dioxide and water so that draining and burning it achieves a reduction by that factor. The total firedamp released per tonne of coal produced and the percentage of this firedamp captured in a drainage system can readily be calculated. It can be shown that firedamp drainage, as recommended by Lyell and Faraday for safety reasons, has the additional benefit of reducing the contribution to global warming from the use of coal for energy production by around 10 per cent.

The Faraday lamp

The second recommendation of Lyell and Faraday in their report on the Haswell explosion related to the monitoring of the state of the atmosphere in the goaf. It will be recalled that the firedamp, because of its lightness, would be at the roof and therefore inaccessible to detection using the Davy lamp. Lyell and Faraday proposed that gas should be drawn from the cavity through a copper tube by an air-pump syringe. A bladder would then be screwed to the syringe, filled with the gas and 'the bladder being carried away to a safe part of the mine, could easily have the character of its contents examined by a Davy lamp'. As with the proposal for firedamp drainage, this suggestion seems to have been lost for many years until reinvented by Sir William Garforth around the turn of this century. The idea of drawing a sample of firedamp from high level into a bladder and injecting it into a Davy lamp came to be adopted. The particular design of lamp which would permit this to be done became known as the Garforth lamp (Fig. 8). It was used in mines until recent times and the idea was adapted also to the pellistor-based instruments described earlier. So the Garforth lamp should in truth be described as the Faraday lamp, to take its place in mining history alongside the Davy lamp.

Coal dust in great colliery explosions

Lyell and Faraday's remit was to ascertain the cause of the Haswell explosion and the reasons for the large number of fatalities despite the fact that the airways in the mine were comparatively free of firedamp even a short time before the explosion, as evidenced at the inquest. Most

Fig. 8 The Garforth lamp.

of the deaths were from poisoning by carbon monoxide or 'chokedamp', a product of incomplete combustion. Lyell and Faraday concluded that this was present to a much larger extent than would have been occasioned by the firing of such firedamp as was normally present in the airways. They advanced the proposition that a large volume of firedamp had been released from the goaf by a roof fall but, in addition, and of crucial importance, they identified the burning of coal dust as a factor. Their own words provide the explanation:

> In considering the extent of the fire for the moment of explosion, it is not to be supposed that firedamp is its only fuel; the coal-dust swept by the rush of wind and flame from the floor, roof, and walls of the works would instantly take fire and burn, if there were oxygen enough in the air present to support its combustion; and we found the dust adhering to the face of

the pillars, props, and walls in the direction of and on the side towards the explosion, increasing gradually to a certain distance as we neared the place of ignition. This deposit was, in some parts, half an inch, and in others almost an inch thick; it adhered together in a friable coked state; when examined with the glass it presented the fused round form of burnt coal-dust and when examined chemically and compared with the coal itself reduced to powder, was found deprived of the greater portion of the bitumen, and in some cases entirely destitute of it. There is every reason to believe that much coal-gas was made from this dust in the very air itself of the mine by the flame of the firedamp, which raised and swept it along; and much of the carbon of this dust remained unburnt only for want of air.

Figure 9 shows coked coaldust on wooden props after an explosion at Kames colliery, Ayrshire in 1955 with an appearance much as described by Lyell and Faraday.

Lyell and Faraday concluded that 'on consideration of the character of the goafs, as reservoirs of gaseous fuel, and the effect of dust in the mine, we are satisfied that these circumstances fully account for the apparent discrepancy' (in the quantity of chokedamp). The orientation of their

Fig. 9 Coked coal-dust after an explosion.

explanation and this remark implies that their belief was that the continued burning of firedamp was necessary to bring out the contribution of the coal dust. Hence if firedamp explosions could be prevented, the coal dust would not be a hazard in itself. In this they were wrong. The burning of coal dust can develop into a self-propagating explosion, with the blast wave propagating ahead of the flame and raising the dust to feed the flame. A firedamp explosion can initiate a coal-dust explosion which will then continue to propagate in the absence of firedamp.

Lyell and Faraday's words on coal dust have entered mining history as the first, and almost correct, expression of the role of coal dust in great colliery explosions. But Faraday rapidly came to realize that coal dust would take over from firedamp as the agent in propagating an explosion. In one of his lectures to the Royal Institution, he advanced the explanation beyond what was offered in the report on the Haswell explosion. 'The gas, first of all, commences the evil, and then lights up this matter [the dust of coal] of which the whole place consists, floor, roof, walls and every part being composed of it. The fire thus gathers energy and goes on ramifying through the mine. It is only on this principle that we can account for the extraordinary extent of injury by chokedamp, the result of combustion, causing 95 deaths in what was considered a very safe mine.' The ramification referred to by Faraday could be very extensive indeed. After an explosion at Risca colliery, Monmouthshire, in 1881 it was found that flame had filled five miles of galleries. However, Faraday's assertion that the gas commences the evil and thereafter (by implication) is not needed to ramify the fire (i.e. explosion) was not accepted in the absence of direct evidence. The conclusion after every great explosion throughout much of the nineteenth century was that the explosion resulted solely from the firing of a sudden large emission of firedamp.

The evidence in favour of Faraday's belief slowly accumulated, mainly provided by careful and systematic investigations by Professor William Galloway[17,18]. In the early part of his career Galloway was an Inspector of Mines in South Wales and made many observations following explosions. He was unable to reconcile the observations with the hypothesis that the explosions had been due only to the combustion of firedamp. He noted that great explosions only occurred in dry and dusty mines, and never in a wet mine. In connection with an explosion at Llan colliery near Cardiff in December 1875, he made the important point that the dryness and dustiness of a mine depend on the depth and the weather at the surface. As the air descends into a mine it is heated up to the temperature of the workings. (As pointed out earlier, the rock temperature increases progressively with depth.) If the dewpoint of the air is originally higher than the rock temperature, condensation occurs; if it is lower, moisture is absorbed by the air. There will be a certain depth at which the temperature is higher than the dewpoint at the surface at any time of the year and, if there is no water penetration, the workings are necessarily dry at all times. Galloway surmised from his observations that this depth lay somewhere between 120 and 200 m and, in addition, that cold weather could change mines that were usually damp into dry ones while it lasted. The explosion at Llan colliery occurred on 6 December following two weeks of below-freezing weather. Explosions also occurred nearby at New Tredegar and Swaithe collieries on 4 and 6 December, lending further support to Galloway's hypothesis.

This evidence was not sufficient since no one had yet been able to show that air and coal dust alone could propagate an explosion. There was ample evidence that even small concentrations of firedamp, below one per cent in air, were sufficient to render coal dust explosive. Since such concentrations were too low to be detected by the safety lamp, there could be no certainty that explosions propagating in a dusty mine were not doing so as a result of the presence of firedamp below the limit of detection. So long as that situation prevailed there could be no conclusive proof against the firedamp emission explanation of great explosions. The point was of more than theoretical importance, since sudden emissions of firedamp could be said to be inevitable whereas the presence of coal dust was a matter of mining practice and therefore controllable, though at a cost. Galloway became increasingly exasperated with the opponents of the coal dust explosion explanation. He gave vent to his feelings in discussing an explosion at Seaham colliery in 1881, stating 'that the flame branched into every district of the workings; that the same insuperable difficulties were encountered in trying to account for the presence of firedamp; and finally, that here also an outburst of gas from the strata was assumed, with its concomitant stultification of scientific methods of reasoning'.

The controversy was eventually settled by two pieces of evidence. The first was the demonstration by Galloway that the ease of propagation of a dust explosion depends on the scale. The proof that coal dust alone could not propagate an explosion had been derived from experiments at the scale of laboratory test rigs. In 1882, Galloway undertook experiments in an outdoor test rig at Llwynypia colliery, funded by the Royal Society, using a duct 0.6 m square and 38 m long. He showed conclusively for the first time that a coal-dust explosion, initiated by a firedamp explosion, could be self-supporting in air in this large-scale apparatus. With this evidence, the views that he advocated at last began to gain acceptance amongst practical mining men. However, there was still considerable opposition in official circles. A book by two Inspectors of Mines described some of the great explosions in the North of England and, according to Galloway, 'put the case for coal dust in a clear and convincing manner; but their book was quickly suppressed by authority and withdrawn from circulation almost immediately after publication'[18].

Galloway's conclusion that a large-scale apparatus was necessary was an important development. He reasoned that the conditions at the actual scale of a mine roadway would be even more favourable to the ability of the explosion to generate the disturbance necessary to supply itself with fuel. Ever since, research on coal dust explosions has been carried out in large galleries, either underground (as in the US, Germany, and Poland) or on the surface (as at Buxton, Derbyshire). Figure 10 shows a coal-dust explosion emerging from the open end of the Buxton gallery.

The event which more than any other turned the tide of expert opinion in favour of coal dust was the explosion at Altofts colliery near Leeds on 2 October 1886. The colliery was virtually free of firedamp. Although candles were used, no workman had been injured by an explosion of firedamp great or small in the twenty years of operation of the colliery.

Fig. 10 Coal-dust explosion emerging from an experimental mine gallery.

The conclusive proof of the role of coal dust, which had been sought for so long, was provided by an unusual combination of circumstances, subsequently described by Galloway[19]. Everyone who visited the workings after the explosion agreed that the explosion was due to the combustion of coal dust alone in the complete absence of firedamp. The jury at the inquest gave a verdict in accordance with that view, the first time this had been done.

The Coal Mines Regulation Act of 1887 took cognizance of coal dust for the first time as a factor in explosions. The precautions recommended fell short of requirements and explosions continued to occur. After two disastrous explosions in 1890, the Home Secretary—in answer to a Parliamentary Question—invited suggestions for the prevention of explosions. On the following day, Galloway wrote to him recommending the simple expedient of frequent watering of the coal dust in the mine roadways. He gave his letter wide publicity with the result that a Royal Commission on Coal Dust was established. Their report made recommendations regarding watering of the coal dust and the use of explosives which were adopted by the Home Secretary. The need to take precautions was at last fully recognized, nearly sixty years after Faraday had first identified the nature of the problem.

Subsequent developments in the means to prevent coal-dust explosions have included making the coal dust inert by the addition of limestone dust or the collection and removal of the coal dust either manually or with vacuum cleaners. Much effort has also been devoted to the development of means of suppressing an explosion so as to contain its effects. A commonly used method is to suspend trays of limestone dust or water in the roadways. The blast wave tips the contents into the path of the succeeding flame and extinguishes it. An alternative is to use sealed containers of limestone dust or water which are ruptured explosively, the activation being provided by sensors which detect the flame front. Coal-dust explosions in mines are now very rare, the last in this country being at Six Bells colliery, Monmouthshire, in June, 1960.

Outside the mining industry, dust explosions are a recognized hazard in any industry in which flammable dusts are handled. Flammable dusts in this context are not confined to recognized fuels but include such materials as custard powder, sugar, and aluminium. Grain silos are occasionally demolished by explosions of the grain dust. In the design of plant for handling dusts, the usual precaution is to provide means to vent the pressure generated by the explosion. This is done by incorporating panels which either fail mechanically or open in time to prevent the internal pressure rising to the point where the complete plant is demolished.

Concluding remarks

The train of development from Dr Gray's letter of 1815 to Sir Humphrey Davy through to modern times provides a fascinating perspective on the application of science to safety and the quality of life. The 'curious phenomenon' accidentally observed by Davy in his researches on the safety lamp provides an outstanding example of the serendipitous character of scientific investigation. The results of Davy's curiosity-driven science have surfaced in today's generation as the main means for reducing pollution from motor vehicles. Faraday's observations after the Haswell explosion provide an important illustration of the wisdom of what we now call the precautionary principle. The implementation of this principle would have defeated the resistance to practicable measures to avoid coal-dust explosions, delayed until full scientific certainty was finally established 60 years after Faraday first expounded on the problem. The scale of human suffering in the interim was truly horrendous: in the period 1879–80 alone, 474 miners died in four separate disasters. The world of safety truly owes a great debt to Davy and Faraday.

Acknowledgements

I am grateful to many colleagues in the Health and Safety Laboratory, Sheffield and Buxton, for their comments on this paper. In particular I would like to thank Geoffrey Eaton for his able assistance in the development and performance of demonstrations supporting the Discourse and to Dr Bryson Gore of the Royal Institution for additional demonstrations from his stock.

References

1. Sir Humphry Davy, *On the Safety Lamp for Coal Mines with some Researches on Flame*, R. Hunter, London, 1825.
2. Irvin Saxton, *The Mining Engineer*, 1986, **145**, 490.
3. Sir Humphry Davy, *Phil. Trans. Roy. Soc.*, 1817, **107**, 77.
4. M. Castel, *Annales des Mines*, 1881, **7(XX)**, 509.
5. H. Le Chatelier, *Annales des Mines*, 1892, **(II)**, 469.
6. A. V. Jones and R. P. Tarkenter, *Electrical Technology in Mining*, Peter Peregrinus Ltd, London, 1992.
7. Sir Humphry Davy, *Phil. Trans. Roy. Soc.*, 1816, **106**, 1.
8. H. Phillips, *Combustion and Flame*, 1972, **19**, 187.
9. Sir Humphry Davy, *Phil. Trans. Roy. Soc.*, 1817, **107**, 45.
10. B. Lewis and G. von Elbe, *Combustion, Flames and Explosions of Gases*, third edition, Academic Press Inc., London, 1987.

11. British Standard 1259, *Intrinsically Safe Electrical Apparatus and Circuits*, 1945.
12. Sir James Lighthill, *Proc. Roy. Soc.*, 1986, **A407**, 35.
13. H. Lloyd and E. M. Guenault, *The Use of Break-Flash Apparatus No. 3 for Intrinsic Safety Testing*, Research Report No. 33, Safety in Mines Research Establishment, Sheffield, 1951.
14. D. W. Widginton, *Ignition of Methane by Electrical Discharges*, Research Report No. 240, Safety in Mines Research Establishment, Sheffield, 1966.
15. J. Adler, F.B. Carleton, and F.J. Weinberg, *Proc. Roy. Soc.*, 1993, **A440**, 443.
16. C. Lyell and M. Faraday, *Report on the Explosion at the Haswell Collieries*, in *Reports from Commissioners* Vol. **XVI**, Committee of Privy Council for Trade, London, 1845.
17. W. Galloway, I. *Proc. Roy. Soc.*, 1876, **XXIV**, 354; II. *Proc. Roy. Soc.*, 1879, **XXVIII**, 410; III. *Proc. Roy. Soc.*, 1882, **XXXIII**, 490; IV. *Proc. Roy. Soc.*, 1882, **XXXIII**, 437; V. *Proc. Roy. Soc.*, 1884, **XXXVII**, 42.
18. W. Galloway, *Great Colliery Explosions and Their Means of Prevention*, Colliery Guardian Co. Ltd, London, 1914.
19. W. Galloway, *Proc. Roy. Soc.*, 1887, **XLII**, 174.
20. P. Shead, *Mining Technology*, 1996, **78**, 39.

JAMES McQUAID

Born 1939, educated at University College, Dublin, and Jesus College, Cambridge. Worked in industry (British Nylon Spinners Ltd and ICI) and the Scientific Civil Service, becoming Research Director of HSE in 1985, and in 1992 Director of the Strategy and General Division and Chief Scientist. He has researched extensively on industrial safety, and received the Bill Doyle Award of the American Institute of Chemical Engineers and the Franklin Medal of the Institution of Chemical Engineers. He is a past President of the Midland Institute of Mining Engineers and of the Sheffield Trades Historical Society, and is a Fellow and Member of Council of the Royal Academy of Engineering.

THE ROYAL INSTITUTION

The Royal Institution of Great Britain was founded in 1799 by Benjamin Thompson, Count Rumford. It has occupied the same premises for nearly 200 years and, in that time, a truly astounding series of scientific discoveries has been made within its walls. Rumford himself was an early and effective exponent of energy conservation. Thomas Young established the wave theory of light; Humphry Davy isolated the first alkali and alkaline earth metals, and invented the miners' lamp; Tyndall explained the flow of glaciers and was the first to measure the absorption and radiation of heat by gases and vapours; Dewar liquefied hydrogen and gave the world the vacuum flask; all who wished to learn the new science of X-ray crystallography that W. H. Bragg and his son had discovered came to the Royal Institution, while W. L. Bragg, a generation later, promoted the application of the same science to the unravelling of the structure of proteins. In the recent past the research concentrated on photochemistry under the leadership of Professor Sir George (now Lord) Porter, while the current focus of the research work is the exploration of the properties of complex materials.

Towering over all else is the work of Michael Faraday, the London bookbinder who became one of the world's greatest scientists. Faraday's discovery of electromagnetic induction laid the foundation of today's electrical industries. His magnetic laboratory, where many of his most important discoveries were made, was restored in 1972 to the form it was known to have had in 1854. A newly created museum, adjacent to the laboratory, houses a unique collection of original apparatus arranged to illustrate the more important aspects of Faraday's immense contribution to the advancement of science in his fifty years at the Royal Institution.

Why the Royal Institution is Unique

It provides the only forum in London where non-specialists may meet the leading scientists of our time and hear their latest discoveries explained in everyday language.

It is the only Society that is actively engaged in research, and provides

lectures covering all aspects of science and technology, with membership open to all.

It houses the only independent research laboratory in London's West
End (and one of the few in Britain)—the Davy Faraday Research Laboratory.

What the Royal Institution Does for Young Scientists

The Royal Institution has an extensive programme of scientific activities
designed to inform and inspire young people. This programme includes
lectures for primary and secondary school children, sixth form con-
ferences, Computational Science Seminars for sixth-formers and Mathematics Masterclasses for 12-13 year-old children.

What the Royal Institution Offers to its Members

Programmes, each term, of activities including summaries of the
    Discourses; synopses of the Christmas Lectures and annual Record.
Evening Discourses and an associated exhibition to which guests may be
    invited.
An annual volume of the *Proceedings of the Royal Institution of Great
    Britain* containing accounts of Discourses.
Christmas Lectures to which children may be introduced.
Meetings such as the RI Discussion Evenings; Seminars of the Royal
    Institution Centre for the History of Science and Technology, and
    other specialist research discussions.
Use of the Libraries and borrowing of the books. The Library is open
    from 9 a.m. to 9 p.m. on weekdays.
Use of the Conversation Room for social purposes.
Access to the Faraday Laboratory and Museum for themselves and
    guests. Invitations to debates on matters of current concern, evening
    parties and lectures marking special scientific occasions.
Royal Institution publications at privileged rates.
Group visits to various scientific, historical, and other institutions of
    interest.

Evening Discourses

The Evening Discourses have been given regularly since 1826. They
cover all aspects of science and technology (with regular ventures into
the arts) in a form suitable for the interested layman, and many scientists

use them to keep in touch with fields other than their own. An exhibition, on a subject relating to the Discourse, is arranged each evening, and light refreshments are available after the lecture.

Christmas Lectures

Faraday introduced a series of six Christmas Lectures for children in 1826. These are still given annually, but today they reach a much wider audience through television. Titles have included: 'The Languages of Animals' by David Attenborough, 'The Natural History of a Sunbeam' by Sir George Porter, 'The Planets' by Carl Sagan and 'Exploring Music' by Charles Taylor.

The Library

The Royal Institution library reflects the functions and the activities of the RI. The subject coverage is science, its history, its role in society including education, and its interaction with religion, literature, and the arts. The emphasis is on the popular science books, the history of science, and the research monographs of interest to the research group in the Davy Faraday Research Laboratories.

It is probably the only library of its kind specializing in the public understanding of science, that is science for the non-specialist. It also has a junior section.

Schools' Lectures

Extending the policy of bringing science to children, the Royal Institution provides lectures throughout the year for school children of various ages, ranging from primary to sixth-form groups. These lectures, attended by thousands, play a vital part in stimulating an interest in science by means of demonstrations, many of which could not be performed in schools.

Seminars, Masterclasses, and Primary Schools' Lectures

In addition to educational activities within the Royal Institution, there is an expanding external programme of activities which are organized at venues throughout the UK. These include a range of seminars and master classes in the areas of mathematics, technology and, most recently, computational science. Lectures aimed at the 8-10 year-old age group are also an increasing component of our external activities.

Teachers' Workshops

Lectures to younger children are commonly accompanied by workshops for teachers which aim to explain, illustrate, and amplify the scientific principles demonstrated by the lecture.

Membership of the Royal Institution

Member

The Royal Institution welcomes all who are interested in science, no special scientific qualification being required. By becoming a Member of the Royal Institution an individual not only derives a great deal of personal benefit and enjoyment but also the satisfaction of helping to support the unique contribution made to our society by the Royal Institution.

Family Associate Subscriber

A Member may nominate one member of his or her family residing at the same address, and not being under the age of 16 (there is no upper age limit), to be a Family Associate Subscriber. Family Associate Subscribers can attend the Evening Discourses and other lectures, and use the Libraries.

Associate Subscriber

Any person between the ages of 16 and 27 may become an Associate Subscriber. Associate Subscribers can attend the Evening Discourses and other lectures, and use the Libraries.

Junior Associate

Any person between the ages of 11 and 15 may become a Junior Associate. Junior Associates can attend the Christmas Lectures and other functions, and use the Libraries. There are also visits organized during Easter and Summer vacations.

Corporate Subscriber

Companies, firms and other bodies are invited to support the work of the Royal Institution by becoming Corporate Subscribers; such organizations make a very valuable contribution to the income of the Institution and so endorse its value to the community. Two representatives may attend the Evening Discourses and other lectures, and may use the Libraries.

College Corporate Subscriber

Senior educational establishments may become College Corporate Subscribers; this entitles two representatives to attend the Evening Discourses and other lectures, and to use the Libraries.

School Subscriber

Schools and Colleges of Education may become School Subscribers; this entitles two members of staff to attend the Evening Discourses and other lectures, and to use the Libraries.

Membership forms can be obtained from: The Membership Secretary, The Royal Institution, 21 Albemarle Street, London W1X 4BS. Telephone: 0171 409 2992. Fax: 0171 629 3569